ADULT EDUCATION IN AGRICULTURE

ADULT EDUCATION IN AGRICULTURE

Ralph E. Bender
Clarence J. Cunningham
Robert W. McCormick
Willard H. Wolf
Ralph J. Woodin
All of The Ohio State University

CHARLES E. MERRILL PUBLISHING COMPANY
A Bell & Howell Company
Columbus, Ohio

THE MERRILL SERIES IN CAREER PROGRAMS

International Standard Book Number: 0-675-09236-1

Library of Congress Catalog Card Number: 71-168863

1 2 3 4 5 6 7 8 — 77 76 75 74 73 72

Printed in the United States of America

THE MERRILL SERIES IN CAREER PROGRAMS

In recent years our nation has literally rediscovered education. Concurrently, many nations are considering educational programs in revolutionary terms. They now realize that education is the responsible link between social needs and social improvement. While traditionally Americans have been committed to the ideal of the optimal development of each individual, there is increased public appreciation and support of the values and benefits of education in general, and vocational and technical education in particular. With occupational education's demonstrated capacity to contribute to economic growth and national well being, it is only natural that it has been given increased prominence and importance in this educational climate.

With the increased recognition that the true resources of a nation are its human resources, occupational education programs are considered a form of investment in human capital—an investment which provides comparatively high returns to both the individual and society.

The Merrill Series in Career Programs is designed to provide a broad range of educational materials to assist members of the profession in providing effective and efficient programs of occupational education which contribute to an individual's becoming both a contributing economic producer and a responsible member of society.

The series and its sub-series do not have a singular position or philosophy concerning the problems and alternatives in providing the broad range of offerings needed to prepare the nation's work force. Rather, authors are encouraged to develop and support independent positions and alternative strategies. A wide range of educational and occupational experiences and perspectives have been brought to bear through the Merrill Series in Career Programs National Editorial Board. These experiences, coupled with those of the authors, assure useful publications. I believe that this title, along with others in the series, will provide major assistance in further developing and extending viable educational programs to assist youth and adults in preparing for and furthering their careers.

Robert E. Taylor
Editorial Director
Series in Career Programs

FOREWORD

Most practitioners in agricultural education accept without question the necessity for adult education in agriculture. Few teachers or county agents will argue with statements like the following:

—To remain a productive citizen, an individual must be engaged actively in continuous learning throughout his lifetime.

—The teacher's degree of commitment to adult education determines his effectiveness in developing relevant and meaningful programs.

—Much of the success or failure in education depends upon the quality of teaching; adult education is no exception.

—Adults will participate in educational programs if the programs are designed to meet their needs to improve their businesses or to be more successful citizens.

—Most successful programs make use of planning or advisory committees to meet the interests and needs of those being served.

—Generally the adult program in agriculture is not the teacher's program; it is shared with the students.

Subscribing to these ideas is one thing; however, making the ideas operational is frequently one of the practitioner's most difficult problems. The authors' major aim is to help those who are or will be teaching adults to bridge this gap between principle and practice. They base their discussion on generally accepted principles of adult education, then proceed in a direct and concise manner to describe and illustrate how those who teach adults can plan and conduct agricultural education programs which exemplify these principles.

In the introductory chapters the authors propose a rationale for adult education and describe the setting conducive to effective programs. Recommendations for identifying and enrolling clientele are followed by procedures for planning adult education programs. The authors devote three chapters to the task of teaching, including the basis for effective teaching, classroom methods, and techniques which supplement classroom instruction. The use of mass media and the value of organizations for adult enrollees, particularly Young Farmers' Associations, are unique features of the book. The concluding chapter outlines evaluation procedures that make evaluation an integral part of the teaching-learning process.

The authors' experiences in adult education are both varied and substantial. All have taught farmers and agribusinessmen as high school teachers of agriculture or county agents. All are now teachers of adult educators; one author directs a university-wide program of continuing education.

The book is written specifically for teachers of adults in public school vocational-technical education and the Cooperative Extension Service. Cooperative Extension personnel at the county and area levels and teachers of agriculture in high schools, area vocational centers, technical institutes, and community colleges will find the book helpful. It is particularly suitable for advanced undergraduates and graduate students enrolled in adult education courses in universities. I propose that those who read it will agree that the authors have been successful in their goal to help close the "gap between commitment to lifelong learning and an operational commitment to adult education."

October, 1971 *J. Robert Warmbrod*

PREFACE

The primary purpose of this book is to aid in the development of adult education programs in agriculture. Almost everyone accepts the fact that some training or education is necessary to keep up to date. The continuous application of science and technology produces new products and methods—a new situation to which one must adjust. Farmers and other agriculturalists are no exception.

Even though many adults have been served through programs conducted by teachers of vocational agriculture and the Cooperative Extension Service, much needs to be done. There are countless opportunities and responsibilities that should be met by these agencies through more specialized programs for farmers. Others that should be served include those engaged in agribusiness, horticulture, mechanics, food processing, conservation, forestry, and environmental protection.

The key person in the development of adult programs in agriculture is the teacher of agriculture or extension agent. Usually the teacher recognizes the need for such instruction, but he needs to be sold on the idea well enough to cause him to give the development of such a program the priority that is necessary. He should be convinced that farmers or other agriculturalists have problems, that they want help, and that they will support and participate in a program. His further reasoning should add to his conviction that he is capable of providing the leadership necessary for a successful program. Without question, however, he needs to make use of the experience of others who have been effective in adult teaching. He must recognize that there are procedures and practices that can be followed in planning, conducting, and evaluating the program such as those included in this book.

The authors sought to identify the rationale and setting for programs, to suggest methodology and techniques to make the teaching effective, and to set forth a guide for securing the kind of clientele who should be served. Because of the scope of the material presented, responsibility for information was divided among the authors by chapter, in the following manner: Chapters 1 and 2, Robert W. McCormick; Chapters 3 and 9, Ralph E. Bender; Chapters 4 and 5, Clarence J. Cunningham; Chapters 6 and 7, Willard H. Wolf; and Chapters 8 and 10, Ralph J. Woodin.

The authors took the position that adult programs are not for the teacher but for the students—the clientele to be served. Therefore, the students should be actively involved in planning and conducting the program. There is no question that involvement of the people will create a more meaningful learning situation. The authors are hopeful that many teachers of adults, prospective teachers of adults, members of advisory groups, and others will be interested in becoming involved with this book. They will appreciate constructive criticism and suggestions to make it a more helpful source of information.

Ralph E. Bender
Clarence J. Cunningham
Robert W. McCormick
Willard H. Wolf
Ralph J. Woodin

CONTENTS

Chapter 1 RATIONALE FOR ADULT EDUCATION IN AGRICULTURE

The increasing complexity of modern life, coupled with our increased standard of educational expectation, has clearly demonstrated that education must be a lifelong process. Professional educators and the general public are accepting more firmly each year the concept of continuous or lifelong learning. While this acceptance is not yet complete, it is inevitable if our educational institutions are to remain viable. An individual, to remain a productive citizen, must not only accept the concept, but must be engaged actively in the process of continuous learning throughout his lifetime.

Gardner (2, p. 4) indicates that "We have abandoned the idea that education is something which takes place in a block of time between six and eighteen (or twenty-two) years of age. It is lifelong. We have abandoned the idea that education is something that can occur only in a classroom."

One rarely finds an agricultural educator or any other educator who will overtly deny the validity of the concept of lifelong learning or the importance of adult education. However, the gap between an *intellectual* commitment to lifelong learning and an *operational* commitment to adult education is still rather substantial. Most agricultural educators, in any facet of agricultural education, tend to begin their careers primarily in youth education and become involved in adult education somewhat reluctantly, if at all. There are many possible reasons why this phenomenon occurs.

Certainly, the formal educational establishment devotes substantially more effort, attention and fiscal and human resources to the education of youth than to the education of adults. Often, young educators are reluctant to enter adult education because they are concerned about their ability to teach adults. They are concerned that their lack of experience in the field may cause adult students to question and probe deeply into the subject matter areas under discussion, and they may see this as a threatening situation. It would seem more rational to interpret the adults' active participation as a deep concern, a deep interest and a desire to make use of the subject matter content.

This book is intended to aid in the transition from an *intellectual acceptance* of the desirability of adult education to an *operational commitment* to develop effective teaching-learning situations for adults. It seems desirable, in studying the need for developing strong, dynamic, continuous adult education programs in agriculture, to analyze relevant current and future environmental conditions. A clear understanding of these environmental factors should, in fact, indicate to all agricultural educators that adult education in agriculture is not only desirable, but is imperative (3, p. 8).

Societal Factors Affecting Adult Education

Agricultural education has pioneered the effective development of adult education programs through significant events which have occurred during the past one hundred years.

AGRICULTURAL LEGISLATION

The passage of the Morrill Act of 1862, establishing the Land-Grant Colleges, combined with the establishment of the United States Department of Agriculture in the same year, was a significant precursor to the thrust in adult agricultural education. In addition to peripheral factors causing the establishment of the Land-Grant Colleges and Universities, the emphasis on education in agriculture and the mechanical arts was designed to provide improved social and economic conditions for the rural population. The Hatch Act of 1887, which established the Agricultural Experiment Stations, had an even deeper impact on the need for and the development of adult education in agriculture, by making it possible to apply basic scientific findings to the solution of "real world" agricultural problems. Meaningful adult education programs in agriculture —either through the public schools, the Cooperative Extension Service or through any other organizational mechanism—would not have been possible without the firm foundation of the research and development activities of the Agricultural Experiment Stations.

The Smith-Lever Act of 1914, which established the Cooperative Extension Service in a cooperative relationship with the federal, state and local governments, and the Smith-Hughes Act of 1917, which provided for vocational education in agriculture in the public schools, were a more specific response to the need for adult education in agriculture. The apparent rationale for the establishment of the Cooperative Extension Service was a desire to provide, through a mechanism of the Land-Grant College, an adult educational program in agriculture and home economics. The Smith-Hughes Act, though it did not exclude adult education, was directed primarily towards the education of young people in the public schools. Clearly, the educators involved in both of these institutions have substantively determined that emphasis on youth education alone is not sufficient; and further, that youth education cannot be separated from adult education. Perhaps the genius of both of these forms of agricultural education has been their ability to move through the full spectrum of the educational process on all age levels.

THE GENERATION OF NEW KNOWLEDGE

The accelerating rate of growth of knowledge in all fields of endeavor, but particularly in agriculture, created largely by the excellent work of the Agricultural Experiment Stations, created the *raison d'être* of substantive adult education programs in agriculture. However, the generation of new knowledge in itself is not particularly valuable to the agricultural producer or any of the other personnel in the agri-business spectrum; it is the application of this new knowledge and the integration of these new findings into the operational aspects of agriculture which has its impact on society. In fact, new knowledge currently is being generated, through the Agricultural Experiment Stations, Colleges of Agriculture and Home Economics, and the U.S. Department of Agriculture, at an accelerated pace. One can only anticipate that new knowledge will continue to be generated at an increasing rate as we move into the future. Thus, if there was validity for establishing a mechanism for educating adults in the past, it is becoming even more imperative that we accelerate our adult education activities now and in the future. It seems equally important that this increase in activity in adult education transcend the pitfall of "more-of-the-same" activities. To be effective, agricultural educators must be concerned with innovations in programming approaches as well as with the adoption of new agricultural practices in serving their clientele (4, p. 255).

SOCIAL CHANGE

Dramatic changes in our society, such as the shift of population from a rural to an urban setting, have been occurring at an accelerated rate during the past few decades—so much so that the distinction between rural and urban is no longer sharp and clear. To say that adult education in agriculture is confined to the rural areas is, in a sense, to deny reality. Adult education in agriculture is needed in extremely urban settings, and agricultural producers' need for adult education cannot be met by dealing exclusively with the subject matter content traditionally perceived as being agricultural in nature. The agricultural producer and personnel engaged in agri-business activities are concerned with, or should be made aware of, the problems of the inner city, since these problems affect their way of doing business and their way of life very substantially. Persons engaged in agriculture must be concerned about all facets of United States domestic and foreign policy. Dramatic changes in these areas challenge the agricultural educator to define clearly the problem areas needing attention in adult education and to utilize effectively in the educational process resource personnel or resource materials which are appropriate to the solution of the problems (5, pp. 102-107).

Adult education in agriculture has parameters; it does not purport to serve everyone, solve all of the problems of society and be all things to all people. However, adult education in agriculture does transcend education exclusively for the agricultural producers or the farmer.

SPECIALIZATION

One potential determinate of the limit of adult education in agriculture is the specialization of subject matter implied by the term "agriculture." This specialization ranges from new developments in agricultural production practices (i.e., soil fertility, animal nutrition, and floriculture production) through the agri-business spectrum.

For example, using the subject matter based approach, significant programs of adult education in agriculture can be conducted with enrollees from suburbia who wish to learn about developing more effective lawn and landscaping procedures. A recent publication of an agricultural research and development center (6) presents a comprehensive study of managerial leadership in supermarkets.

While the subject matter approach provides a fairly logical delineation of the activities of adult education in agriculture, it does not provide the complete rationale for adult education activities. The acceptance of the practicing agricultural educator by the people in a local community or county (particularly a rural community) enables him to extend his adult education activities into subject matter areas outside the field of agriculture. Because the *clientele groups* have confidence in the agricultural educator, they may ask him to provide educational activities in such diverse problem areas as community health, zoning regulations, or education for local government officials. These requests illustrate a unique phenomenon of adult education in agriculture, which is the *deep involvement* of the potential participants in the planning of their own educational experience, as well as the intensive interaction of the educator and his "clientele" within the community. The involvement of prospective participants in program development is desirable in all forms of adult education.

Programs which require expertise which is clearly beyond the knowledge of the agricultural educator will be meaningful only when he is able to obtain the resources required from the university, from other educational agencies or from appropriate sources in the community.

The increased specialization of functions of all citizens in our society, and particularly those in the field of agriculture, is a factor which affects not only the nature but the scope of adult education activities in agriculture.

One needs only to look at the production aspect of the field of agriculture to see very clearly the increasing specialization of activity. The small firms engaged in "general agricultural production" have become a marginal kind of activity in some areas, and have ceased to exist in others, and one can safely predict for the future that they are likely to disappear from the scene. This does not suggest that there is no need for us to be concerned about adult education in agriculture in the future, but only that the nature and the level of specialization of our programs must be substantively altered as we move into the future. We know that it is not uncommon for an agricultural producer to have a quarter of a million dollars invested in his enterprise. Clearly, highly sophisticated managerial expertise is required to operate this type of business, and producers in this category more and more want the latest information about new developments in their specialized fields of expertise.

INCREASED EDUCATIONAL LEVEL

The increased level of formal education for all of our citizenry will have a significant impact on adult education programs in the future; as the formal level of education increases, there is a desire for more variety in adult education programs. The belief prevails that the more education an individual has attained, the more educational opportunities he is likely to seek. This suggests that adult education programs in agriculture not only must serve the specific needs of the clientele, but also must be presented in such a way that they will be attractive to the clientele.

The utility of adult education in agriculture, not only for agricultural producers but for the entire population of the country, has been a very salable commodity in the past and should continue in the future. For example, the adoption of new practices in production agriculture and in the agri-business field often enhances the efficiency of food production and marketing. The standard of living in the United States would not be so high if the production of the necessary food and fiber still required the efforts of 80 percent of the country's population. Clearly, the efficiency of agricultural production has been substantively increased through agricultural research and adult education in agriculture during the last century. The efficiencies thus created have released people from the requirement of producing food and fiber so that they can engage in those service and consumer product activities which enhance our standard of living. The demands for increased efficiency in the future will place increasingly strong demands upon adult education programs in agriculture.

The Climate for Adult Education

While the desirability and the need for adult education have long been recognized, the general perception of educational institutions, the federal government and the general public has only recently been directed toward answering this need. This new perception creates a much more desirable climate which widens the scope of adult education to include all levels from basic or literacy education to continuing education for professional people.

Although adult education as a division of the educational establishment has not yet reached the level attributed to elementary, secondary and higher formal education, there is a distinct movement in that direction. Through such significant legislation as the Vocational Education Act of 1963, the Elementary and Secondary Education Act, the Vocational Education Amendments of 1968, and Title I of the Higher Education Act of 1965, the federal government has provided funds for adult and continuing education.

However, professional educators must truly believe that adult education is more than a marginal operation and must commit themselves to demonstrating that the importance of adult education is equal to that of youth education before they can expect prospective participants in adult education to perceive it in that fashion.

The degree of institutional commitment to adult education seems to parallel the level of commitment of the professional educators in the field, and thus it varies from institution to institution. However, some institutions are indicating their increased commitment by providing more "release time" to vocational agriculture teachers engaged in adult education activities. Clearly, only when an institution is prepared to provide adequate staff resources will adult education activities have a significant impact upon that institution.

While there has been a tendency in fields outside of agriculture to consider adult education as a part-time or overload type of activity for the "regular" teachers in the system, there is some evidence that institutions are moving from this position. Even in colleges and universities there is still a strong tendency to consider continuing education or adult education activities as "nice to do" activities as long as they do not interfere with the main function of the institution, which is formal degree education (1, p. 88). Fortunately, because of the strong background and tradition of adult education in agriculture, this concept is less prevalent at all levels in that field than it is in some others. Agricultural education leaders have recognized that adult education activities provide an opportunity for instructors of youth to become involved in the problems of the field and to serve adults in the community directly, and that these experiences often enhance their formal teaching activities.

It seems no accident that the public funds made available for adult education in agriculture far exceed funds available for adult education in other areas of expertise. The Cooperative Extension Services in Agriculture and Home Economics and Vocational Agriculture have been traditionally committed to providing public funds for the development of educational programs to assist in solving problems. If these funds are to continue to flow, agricultural teachers and administrators must be able to demonstrate that these funds can be used effectively by being innovative in their approaches.

The Institutional Scope of Adult Education

The establishment of the Land-Grant Colleges and the Agricultural Experiment Stations, followed by the development of the Cooperative Extension Service and Vocational Agriculture, laid the foundation for adult education in agriculture today.

Vocational education in agriculture has traditionally maintained an adult education component in the total program, but until fairly recently, this component has been manifested in a series of adult or young farmer classes held in the evening during the fall and winter months. These courses were primarily geared to the problems and the adoption of new practices in production agriculture. A vocational agriculture teacher's involvement in an adult education program has been largely a matter of individual choice. If we believe that adult education is a new imperative for our time, vocational agriculture teachers must place an increased emphasis upon adult education.

The development of community colleges, technical schools and technical institutes at both the public school and higher education levels provides new opportunities for adult education in agriculture. The adult education component of these schools will ultimately be a very strong force in the field of adult education in agriculture.

The Land-Grant Colleges and Universities have, by definition, accepted a strong commitment in adult education through their Cooperative Extension Service in Agriculture and Home Economics. However, there is increasing evidence that if the Cooperative Extension Service is to legitimately fulfill its mission as an arm of higher education, it must deal with more sophisticated subject matter in more and more specialized areas and at the same time, it must provide an integrating base in the management area of agriculture.

Coupled with the Cooperative Extension Service at the college and university level are the on-campus short courses, seminars and conferences. These methods are expected to become so important with additional advances in communication, the new electronic media and improved transportation, that they will be used for conducting much of the highly specialized activities in agriculture geared towards highly specialized clientele.

The many diverse agencies involved in agricultural education must engage in cooperative planning and activities to utilize optimally the resources available to them. As the recipients of a large proportion of public funds for adult education, agricultural educators must accept the responsibility for being certain that these various agencies avoid duplication of effort and that they serve the full spectrum of adult agricultural education.

The Future of Adult Education in Agriculture

The future of adult agricultural education appears to be quite bright. It is, of course, contingent upon meeting the participants' needs in appropriate ways, and thus evaluation of outcomes in terms of dollars and cents becomes difficult. However, a measure of the adoption of new practices in the field, of educational change among the participants in the program, and of the percentage of involvement of potential candidates in the program, all contribute to a measure of the effectiveness of the program. Since participation in adult education is essentially voluntary, the agricultural educator is challenged to develop programs that are relevant to today's problems in agriculture and that assist the participants in making appropriate decisions about their enterprises.

Clearly, one of the benefits from adult education accrues to the educational institution when its image is improved through serving adults in meaningful ways through educational programs. In a real sense, the adult education program is an outreach of the institution and when its programs are effective, it will create even further support for local, state or federal appropriations for educational activities that go beyond the adult education programs. It seems incumbent upon agricultural educators who are committed to adult education to promote its adoption at various local institutions.

REFERENCES

1. Birenbaum, William M., *Overlive.* New York: Dell Publishing Co., Inc., 1969.
2. Gardner, John W., *Campus, 1980.* New York: The Delacorte Press, 1968.
3. Jensen, Gale, *et al., Audit Education: Outlines of an Emerging Field of University Study.* Washington: Adult Education Association of the U.S.A., 1964.
4. Rogers, Everett M., *Diffusion of Innovations.* New York: The Free Press of Glencoe, 1962.
5. Shannon, Theodore J., and Schoenfeld, Clarence A., *University Extension.* New York: The Center for Applied Research in Education, Inc., 1964.
6. Trieb, S. E., and Marion, B. W., "Managerial Leadership: Its Effect on Human Resources in Supermarkets," *Research Bulletin 1027.* Wooster: Ohio Agricultural Research and Development Center, August, 1969.

Chapter 2 THE SETTING FOR EFFECTIVE PROGRAMS

We have discussed in the previous chapter some of the general environmental factors that an agricultural educator needs to consider in developing adult education programs; in this chapter we will focus upon his specific setting in the field.

This discussion is addressed to the agricultural educator who is developing adult education programs at the community, county or area level because he interacts daily with his prospective participants, and thus is certainly one of the major factors contributing to the success of adult education programs in agriculture. The validity of the concept of the local educator or "field staff" educator is evidenced by the fact that it has been utilized by other adult education programs, both in the United States and in the developing countries. For example, the concept of the "Village Level Worker" in the Community Development Program in India can be traced directly to the success of the county extension agent in the Cooperative Extension Service in the United States.

Due to the efficacy of these field staff personnel, a number of states have changed the organizational structure of University Extension organizations so that the field staff in the counties serves not only agriculture but also draws upon expertise from other areas of the Land-Grant University (2). In addition, the Technical and Business Services program, established by the State Technical Services Act of 1965, has concentrated a very high proportion of funds upon building a field staff in what is termed "Industrial Extension" (4).

Situational Factors

Every person who attempts to develop adult education programs in agriculture must be aware of the unique situational factors which prevail in his field. The most obvious situational factor is a geographic delineation; he is expected to concentrate his efforts within a community, a school district, a county or some combination of counties. The appointment of area extension agents (or specialists) by the Extension Service is a fairly recent phenomenon, but it has been widely adopted throughout the United States and has had a significant impact upon the role of the local extension educator, and of the vocational agriculture educator as well.

The appointment of area personnel in addition to county extension personnel aids the local extension educator by providing close-at-hand specialized expertise in selected agricultural areas. In addition, many states encourage area extension educators to conduct programs in their specialized areas of expertise *directly* with agri-business personnel and agricultural producers. In a number of states, the Extension Service employs personnel with expertise in non-agricultural disciplines such as political science and social work.[1]

While area extension educators are a recent development, somewhat "within the family," equally significant developments have occurred through mechanisms which are external to either vocational agriculture or extension.

The rapid development of community colleges, technical institutes and area vocational schools provides the potential for conducting programs of adult education in agriculture and for conflicts between these institutions and the local extension educator. There is evidence that this conflict has already been manifested. Community colleges and area technical schools have developed because of local demand. In addition, they enjoy tremendous local support because evidence of their value is visible to the people of the community. In most cases, these institutions have at least one full-time person engaged in the development of adult education programs, and they have a somewhat natural tendency to develop a proprietary "right" regarding a specific geographic boundary. On occasion,

[1]Wisconsin and Missouri provide examples of this development.

these institutions have suggested that it was their role to coordinate *all* adult education programs (including those in agriculture) in "their" territory (2). Thus, the local educator developing an adult education program in agriculture must, at a minimum:

1. Be familiar with the programs of these institutions.
2. Develop appropriate communication procedures between the several agencies and organizations involved.
3. Involve representatives from these institutions and agencies in appropriate planning groups.

A local educator developing adult education programs in agriculture *must* recognize that his are not the only development efforts in the geographic area he defines as "his area." Even in the most rural community or county there are other relevant programs of adult education which will affect him. An imperative first step for any agricultural educator who is initiating or expanding an adult education program is an inventory of the existing local adult education opportunities and an appraisal of their adequacy. In addition, other situational factors, such as economic, social and population factors, must be appraised.

An analysis of the relevant situational factors need not be a five-year study prior to initiating a program; rather, it should be continuous and should be accomplished in concert with other adult educators in the area. While each educator must develop his own program, each program should help to achieve the objectives of the community.

Adult education programs in agriculture cannot be restricted to the areas of production agriculture or farming; the full agri-business spectrum provides possibilities for meaningful programs in many areas. However, in some instances adult education programs for the agri-business complex may be best developed by the joint efforts of several agencies and organizations.

The Teacher—Key to Effective Programs

In spite of the many environmental and situational factors that bear upon adult education programs, in the final analysis it is the teacher who provides the key to effective programs. A teacher must be more than intellectually committed to adult agricultural education; he must be emotionally involved as well, for the degree of his commitment will determine his effectiveness in developing relevant and meaningful programs for his potential clientele. This commitment is not something that is inherent in one's makeup; it is developed by interaction with the adults in the community.

Although many teaching-learning principles apply to teaching both young people and adults, there are substantive differences in teaching the two groups. The most fundamental difference is that adult education cannot be directed toward use in the future; it must relate to the "here and now," and the teacher who fails to recognize this key difference is likely to be ineffective in dealing with adults.

Innovative approaches to programming are an imperative in adult education since they stimulate maximum participation, which helps adults satisfy their need for deep involvement in the educational process. Perhaps one of the significant benefits to be derived from teaching adults is that innovative programs developed through adult education should provide a basis for innovative modifications in the youth education program.

Administrative Support for Adult Education

While effective programs of adult education in agriculture rest heavily on the teacher of adult education, they also need administrative support. More and more the adult educator in agriculture can be assured of strong support from the State Department of Education and from the State Director of the Cooperative Extension Service in Agriculture and Home Economics (3). Commitments to the program have generally been firmly established at these levels. (Administrative support transcends, but includes, financial support for the program.)

In recent years, administrative support for adult education has increased through the development of state and national programs. This has involved the establishment of broad national and state objectives, within which local programs of adult education can be developed. Mathews has described the Cooperative Extension Service as an organization with national objectives which can be adapted to local needs (1, p. 32).

While some teachers and extension agents have referred to the above-mentioned program leadership as government domination, it is clear that such guidelines and objectives serve more as a stimulus to adult education activity than as an impediment to progress in this area. State and national objectives tend to provide a general focus for the program, but still leave wide latitude for the exercise of individual initiative. However, the opportunity to utilize public funds rarely, if ever, exists without a modicum of control. Any other arrangement would clearly reflect fiscal irresponsibility on the part of the government. Complete local control of education is currently, and has been for some time, a myth; however, state and national control is very limited and certainly not unduly restrictive.

Administrative support for adult education at the local level is undoubtedly more varied than at the state level. Since the Cooperative Extension Service in Agriculture and Home Economics is directly affiliated with the Land-Grant University in each state, the locus of concern for support at the local level is somewhat different than in vocational agriculture.

Teachers of vocational agriculture need to build support with the executive head or administrator of the local school and through him, to the local board of education. There is an equal need for building support with the county government officials in Cooperative Extension. Since a significant portion of funds for extension personnel come from the county commissioners or board of supervisors, it is imperative that they support the program fiscally.

Generally, effective advisory committees, made up of lay citizens in the local area, perform a valuable function in building local support. A new county extension agent moving into a county will undoubtedly find some type of an advisory committee in existence; however, in some cases this committee may need to be further developed. A vocational agriculture teacher's new school district may or may not have had an adult education program before; thus, it may be desirable to build support for the program with the local administrator and have him build support with the board of education before organizing a lay committee.

Building Public Acceptance for Adult Education

Unless there is local support and public acceptance of adult education, the program will likely be doomed to failure. The development of effective lay committees and the building of programs on clearly identified needs and interests of the local community will help to enhance the overall public acceptance of the program. The prepackaged adult education program which the local teacher or administrator attempts to sell to the public may find very poor acceptance even though it may be useful and well developed. The appropriate and deep involvement of potential participants in the program is a key factor in building public acceptance.

Adequate communication about the adult education program, once it has been built, developed and conducted, furthers public acceptance for this kind of program. This communication may take several forms, the most effective of which is positive word-of-mouth communication by the current participants in the program. Personal contact between the instructor and the public is a significant and powerful force but is also expensive. However, appropriate use of mass media can also build public acceptance. One must keep in mind that the basic element of building public acceptance is the development of a meaningful and effective program.

Developing Adequate Financing for the Program

There is a reasonable level of federal, state and local financial support for adult education programs in agriculture. However, the adequacy of these levels varies from state to state and county to county. Generally speaking, these public funds provide a base from which one begins to operate. One may add to the financing through other means.

While participant fees have not been assessed traditionally in adult agricultural education, there is a distinct trend in this direction. Adult education in agriculture has not been plagued with the need for self support so evident in other aspects of adult education programs; nevertheless, an effective teacher or administrator of adult education should not overlook the possibility of charging participant fees. Assessment of even a modest amount of participant fees provides additional commitment to the program, and thus enhances rather than detracts from the program in most situations. Clearly, there are some kinds of programs that may have to be completely subsidized. But in the main, some financial input from the participant, even though it may be very minimal, will enhance his commitment to the program and his attendance at the various phases of the program.

The many federal acts which are addressed in part to adult education programs make available a variety of local, state or federal grants or contracts for program development. Before engaging in a potential government contract, an institution or an individual teacher of adult agricultural education should review it in terms of how well it serves the interests of the institution and personnel involved.

In general, it is unwise for a public educational institution to enter into a contract with individual businesses or organizations for specific programs; the opportunity for participation should be available to all those who are interested.

Identifying Appropriate Facilities for Adult Education

A youth education facility is often inadequate for adult education. The following is a list of the necessary physical characteristics of an adult education facility:

(1) The facility must be easily accessible with adequate parking space.
(2) The physical conditions should be such that adults will feel free to participate and interact; comfortable chairs and tables arranged in seminar fashion will usually provide the most satisfactory setting for adult education.
(3) The traditional classroom setting should be avoided in most instances.
(4) Appropriate arrangements need to be made for socialization—perhaps a coffee hour or some kind of refreshments served during the meeting.

In adult education, as well as in youth education, effective learning experiences also take place outside the classroom. Tours, demonstrations and other field trips form a significant element of the appropriate setting for adult education programs in agriculture, and need to be a planned part of the total program rather than being added as the program progresses.

Program Coordination with Other Adult Education Agencies

Teachers of vocational agriculture and the local county extension agents need to interact so that they can build complementary approaches to adult agricultural education and be aware of the program opportunities available through both agencies. This interaction will not cause one agency to dominate the other or to dictate the program format; on the contrary, it will invite cooperation, which will ensure the effective use of available resources.

The need for coordinating adult education programs with other adult education agencies will increase in the future. Clearly, adult educators in agriculture must not restrict their programs to the agricultural producer—the farmer. Viewed strictly from a self-interest perspective, a program restricted to this narrow strata of clientele is faced with the prospect of a continuously declining number of participants. One could anticipate the time, given this narrow program definition and no reduction in adult education personnel, when there might be more agricultural educators than farmers. Quite apart from the matter of self-interest, adult educators in agriculture not only have meaningful programs for the non-farm segment of our population, they also have a distinct contribution for other adult educators. Perhaps the most significant contribution made by adult educators in agriculture has been the process by which they have involved the prospective participant in the development of programs to meet the participant's needs. The concepts, principles and experience of agricultural educators related to this process can make a signal contribution to the entire adult education movement, when and if it is utilized by adult educators in fields outside of agriculture.

REFERENCES

1. Mathews, Joseph, "The Cooperative Extension Service," *Convergence,* Vol. I, No. 4, (Winter 1969).
2. *Proceedings of the Council on Extension, National Association of State Universities and Land-Grant Colleges, November, 1969.* Washington: National Association of State Universities and Land-Grant Colleges, 1970.
3. *A Report of Joint U.S.D.A.-National Association of State Universities and Land-Grant Colleges Extension Study Committee,* "A People and a Spirit." Fort Collins: Colorado State University, November, 1968.
4. U.S. House of Representatives: Testimony of the Committee on Industrial Extension, National Association of State Universities and Land-Grant Colleges before Subcommittee on U.S. Department of Commerce Appropriations, May, 1969.

Chapter 3 EFFECTIVE TEACHING FOR ADULTS

Most adults in the United States have experienced formal education and have arrived at some conclusions about it. What they have found to be interesting and helpful they are willing to promote for themselves and for others. If, however, their educational experience was unsatisfactory, they are hesitant to become involved in another program.

Much of the success or failure in education depends upon the quality of the teaching, and adult education is no exception. Since most adult education programs are voluntary, participating adults demand a suitable teaching and learning experience. In general, they want the following:

1. *Adults want their learning to be useful.* They have many pressing problems which demand solutions in order that they can control and enjoy their changing environment, and they are challenged economically to meet the financial demands of their businesses and their families. In addition to economical problems, farmers and other agriculturalists, like all people, have social concerns. They not only want to make a living, but they want to live a life that is enjoyable and "on par" with the members of other segments of their society. Education should contribute to the solution of such problems.

2. *Adults want to be actively involved.* They want to help decide the kind of program to be conducted and to relate it to their own purposes, so that the program will be meaningful to them if they participate in it. Professional personnel should advise, but they should not assume the entire responsibility of deciding what, when, and how to teach, because adults are much more committed and interested in a program if they take part in planning it.
3. *Adults want competent teachers.* Teachers are key persons in developing and conducting the program. They should have a thorough knowledge of their fields and the ability to relate the technical to the practical. The teaching should be interesting, useful and thought provoking. Television and other programs are enjoyable and worthwhile enough to provide competition for the students' time.
4. *Adults want methods which are effective.* They have limited time to devote to adult education; therefore, they will insist that the methods used be efficient in developing useful knowledge and skills. In most cases, theory should be integrated with practice. Each session should contribute to the achievement of their ultimate goals.
5. *Adults want a desirable social experience.* They will participate in adult education meetings and associated activities only if they feel they are wanted and can be with others who have similar interests and standards. Adults are not a captive audience. They do not have to return to the next meeting.

Adults Can and Want to Learn

More and more adults appreciate the fact that some training or education is necessary to keep up to date. The continuous application of science and technology produces new products and methods—new situations to which one must adjust. Farmers and other agriculturalists are no exception; they cannot farm as they did five years ago and continue to be efficient or successful. They want to learn because they know that new developments demand new adaptations in their situations.

Adults not only *want* to learn, they *can* learn. This fact is demonstrated every day by the provisions being made for adult learning experiences and the action being taken by the individuals on their own.

It seems almost nonsensical to ask the question, "Can adults learn?" In the nineteen-twenties Thorndike proved through his program of research that adults are efficient in learning (25). The research findings as well as observations lead to the conclusion that people participating in adult education are self-motivated. They have goals that they want to meet, and their greater maturity usually is conducive to learning.

> It is of interest that when Thorndike published his book in 1928 on adult learning he compared two age groups, 20-24 and 35 years or more. The latter group had an average age of 42 years. A contemporary study of aging would hardly regard a group with an average age of 42 as old and indeed the dialogue concerning aging and capacity to learn has tended to be shifted to comparisons of adults over 65 with those in their twenties. With the gradual rise in years of educational attainment more adults are aging with a positive attitude toward abstract learning as well as learning of psychomotor skills. One example in recent years of the expectancy of adult capacity for learning was shown by the FAA decision to permit commercial airline pilots to train to shift from flying piston airplanes to jet planes up to the age of 55. The idea of an individual learning to fly such a dramatically new type of plane at the age of 50 would come as quite a shock to students of learning in the days of Thorndike (14, p. 8).

Alfred Krebs, in his book *For More Effective Teaching,* analyzes the problem-solving approach used by teachers of agriculture. He points out some of the characteristics of adults and why teaching them is different from teaching high school students. The major factors necessitating some variation or adaptation of teaching procedures are as follows (13, pp. 143-45):

1. Adults have more control over what they do.
2. Adults are capable of considering broad, complex problems.
3. Adults are capable of more self-direction in their consideration of problems.
4. Adults require less motivation from the teacher. They are far more eager to take advantage of educational programs in vocational agriculture than schools are to provide such opportunities.
5. Adults assume more responsibility for the extent to which conclusions arrived at in class meetings are put into practice. This does not mean that the help and guidance of the teacher in making this application are not required; on the contrary, the on-the-job instruction, where planning relates specifically to existing conditions, is probably the most important phase of the adult education program. In the final analysis, the effectiveness of the adult education program in agriculture will be determined by the effectiveness of the on-the-job instruction.
6. Adults not only assume more of the responsibility for the direction in which the class moves, but also for the soundness of the conclusions drawn at the end of the consideration of a particular problem.

7. Adults provide a much richer source of information and knowledge about agriculture within their own group. There is less need for resource persons outside the group itself. When resource persons are brought in from outside the group, they should be used only in a consultant capacity.
8. Courses for adults are usually publicized in terms of specific course titles. This means that there is a homogeneity of interests since the adult farmer will register for the course which he feels will be of the greatest benefit to him.

As people grow older there are some changes that make learning more difficult. For example, there is a decline in energy reserve and in visual and auditory acuity. This usually develops gradually, beginning in the early twenties and taking a distinct downward trend after the age of fifty. Learning rate and reaction time diminish, and adult habits become more rigid with advances in age. These conditions tend to slow behavioral change, and, in some cases, contribute to more impatience and discouragement.

Many adults lack confidence in themselves as learners, particularly in a group situation. Some of this lack of confidence is associated with unpleasant or unsuccessful learning experiences during earlier years. "Adults learn much less than they might, partly because of a self-underestimation of their power and wisdom and partly because of anxieties that their learning behavior will bring unfavorable criticism" (16, p. 7). In some cases their lack of confidence is due to the fact that they have not participated in a formal learning situation for many years, thus allowing the necessary learning skills they acquired earlier to diminish through disuse. These facts may cause adults to fear becoming involved in an adult program and to be less successful in their early experiences as adult learners. Farmers and other agriculturalists are generally not quite as sensitive as the general public in this regard because they have been accustomed to attending agricultural meetings and activities.

Some adults have difficulty in associating new learnings with previously learned patterns. To accept the new, the adult may need to unlearn a behavior which has been practiced and reinforced many times, and this is usually difficult. Because experience is conducive to forming habits and points of view, adults find the acceptance of new learnings more difficult than younger students.

The number and kind of adult role expectations likewise affect the learning efficiency:

> The typical middle class adult may simultaneously be trying to fulfill the role expectations of husband, bank president, country club

> member, church deacon, and student. Making such an extensive and diversified effort may not only reduce the amount of energy he has available to expend on learning; it may also complicate the task itself. For instance, the results of decisions made during the accomplishment of a given learning task may potentially affect the lives of numerous people with whom the adult learner associates in performing other roles (20, p. 207).

The expectations of these "other people" affect the learner's perception, motivation, and success in learning. Therefore, adult teachers need to understand their students' total responsibilities and limitations when planning the program of instruction, the classroom or group activity, and the individual follow-up of the instruction.

Concepts of Teaching and Learning

An analysis of some basic concepts and principles of teaching and learning is helpful to those who are interested in improving teaching. A good teacher will think about methods as well as outcomes in terms of what he knows to be effective teaching. There is no general agreement concerning the best concepts. The review presented here is in terms of effective teaching of adults; however, the same basic concepts and principles apply generally to teaching other groups.

WHAT IS TEACHING?

There are many definitions of teaching; one might be "imparting of information," as when a teacher lectures and passes on facts or subject matter to the students. In another situation, a teacher guides students through a textbook and/or other references so that they will be able to "hand back" such information. Part of this procedure may include questioning, drilling and testing to learn the facts. These concepts of teaching are not appropriate or inclusive enough to satisfy what is desired or needed in teaching adults.

A more acceptable definition of teaching may be "The guiding of the learning process so that pupils will learn more and better." Teaching is a process of intentionally changing behavior. Hammonds and Lamar support this concept when they point out (11, p. 3), "Teaching . . . implies contemplated learning products—teaching objectives. It is the guidance of learning toward desired ends." In answer to the question, "What does it mean to teach?" Edgar Dale replied, "To teach is to transform by informing, to develop a zest for lifelong learning, to help pupils become students—mature, intelligent learners, architects of an exciting, challenging future. Teaching at its best is a kind of communion, a meeting and merging of minds" (4, p. 170).

The teacher's role is to provide the setting, the materials and the procedures that facilitate learning. In carrying out such a process, he should be concerned with what to teach, when to teach, and how to teach.

Another way of looking at teaching is through a study of the competencies of the teacher. More and more teachers are becoming capable of performing multidimensional tasks rather than only teaching a class in a certain subject matter area. This is true of adult education in agriculture. The adult teacher is one who is able to relate to the learning of the individual student as well as to plan, develop and conduct programs for many groups of diverse backgrounds of training and experience.

The late Dr. Kimball Wiles, Dean of the College of Education at the University of Florida, listed the competencies needed by a teacher (30, p. 261):

1. The ability to relate to the learning of one student; this includes diagnosis and individual instruction.
2. The ability to analyze group development and interaction and perform a leadership role in a group.
3. The ability to communicate both with individuals and groups.
4. Knowledge and skill in a discipline or field. This knowledge should be deep enough to enable the teacher to approach the field from a number of different angles and to inquire into dimensions that he has not hitherto explored.
5. The ability to structure and restructure knowledge. This competency will enable the teacher to choose from his specialization the type of knowledge that is important to a given individual or group and to restructure it so that the individual or group may investigate the knowledge in terms of its own motivation.

WHAT IS LEARNING?

Undoubtedly, a teacher can understand his role only if he has a concept of learning. A workable definition may be "learning is the process by which one, through his own activity, becomes changed in behavior." When a person learns, he feels, thinks and acts differently because his interests, skills, abilities, understandings, attitudes and appreciations have been changed. These changes are as obvious as the differences in physical characteristics among our acquaintances. Thus, by accepting this concept, it is possible to see every educational expérience as contributing to differences in a person. The results of learning may not always be observable in overt actions, but a change in the person nevertheless has been made and has had its effect on his total personality.

According to Bruner (1), the act of learning involves three simultaneous processes, including (1) the acquisition of new information, (2) the transformation or manipulation to make it fit new tasks, and (3) the evaluation or checking to see if it is adequate. Jensen, Liveright and Hallenbeck support this concept in the following (12, p. 30):

> Learning may be defined as "the acquisition of information and the mastery of that intellectual behavior through which facts, ideas, or concepts are manipulated, related, and made available for use." The acquisition of information alone is not learning, even though content is the vehicle through which the desired intellectual behavior is mastered. While some adult experience may successfully accomplish one or the other of these two elements of learning, our concern here is with those experiences that achieve both simultaneously.

It is important to understand that learning is a self-active, personal choice-making process—a person learns what he wants to learn. Professor Weir, University of Pittsburgh, reports that ". . . learning is a process of personal choice making; it is activity in which the learner, through his own experience and out of his own motivations, decides for himself what he is to believe and what he is to do with his life" (28, p. 408). He indicates that the individual himself is the most important circumstance in his own environment. He points out: "Man is continuously restructuring his environment so that it takes on meanings that were not there before" (28, p. 409). Successful teaching is that which brings about growth in the learner and helps him to realize his potential. In this context, the teacher is responsible for helping to identify appropriate goals and providing the materials and experiences that are necessary for their achievement. He is the director of the learning process.

Since all learning is creative, it cannot be a duplication of someone else's effort. Each bit of learning is, to an extent, original; it is significant and not necessarily routine. This is particularly true for adults, since they are enrolled in adult education programs primarily because they want to be. When they select a program, they are helping to decide the kind of behavior change they desire.

People vary greatly in their ability to learn and in the rate at which they learn. They have differences not only in mentality, but also in motivation and capacity. Adult students often will learn much more—and faster—than teachers expect because of their background of experience and motivation. For example, it has been found that persons meagerly trained in reading and simple arithmetic can make great improvement in these fundamentals when they have a reason for doing so.

Learning is a process of integrating and relating new experiences with former experiences. Thus, the more basic training and experience a person has in any given area the better equipped he is to learn further in that area.

In most programs of education, the identification of valuable knowledge is one of the most baffling problems. We must recognize that all learning is not necessarily good; one can learn that which is undesirable for his own welfare and the welfare of society as well as that which is good. Adult teachers may teach some practices that are out of date. For example, promoting the use of weed control chemicals that were good last year, but not appropriate currently, would be easy if one were not aware of current research and new and more effective products.

Learning usually brings satisfaction and rewards. A person feels good when he has learned what he wanted to learn, because learning is prompted by goals for which he is striving. Psychologists have pointed out that learning is more effective in an environment of success. A person who has had a number of failures will be more reluctant to take advantage of what other people may believe to be good opportunities.

Much of the above concept of learning may be summarized in the following quotation (18, p. 232): learning is " . . . an episode in which a motivated individual attempts to adapt his behavior so as to succeed in a situation which he perceives as requiring action to attain a goal."

"If we believe that lifelong learning is a necessity, we must accept self-education as a key goal of learning. We must learn how to learn and develop a zest for learning" (5, p. 1). Edgar Dale points out that each learner must get his ignorance organized and decide what he intends to learn. He must learn how to process information, ideas and subject matter. The teacher who accepts this belief should try as much as possible to teach his students how to learn—that is, how to relate and integrate new ideas, facts and practices with the old.

Dale suggests that (5, p. 2):

> . . . If we are to learn—become skillful in meeting new learning situations—we must distinguish between training and education, between imitative reaction and creative interaction. When we train we put emphasis on drill, on imitative behavior, on unrelated specifics, on memorizing, on immediate results, on learning activities which have a low ceiling of growth. We rely only on teacher or university evaluation. We learn how but do not ask why.
>
> When we educate, however, we emphasize creative interaction, learning by discovery, developing usable generalizations from spe-

> cifics. Critical self-evaluation looms large in the process of learning to learn. The student learns to ask both how and why.

These statements have many implications for the successful adult educator who is developing and conducting a program of education that will result in self-direction on the part of the learners. "Learning should not only take us somewhere; it should allow us later to go further more easily" (1, p. 17).

Methods of Teaching

Teachers are interested in methods that can be pursued to promote more and better learning. Methods are the ways and means of relating the teacher and the pupil in learning situations. The only justification for the teacher's presence is that he helps facilitate the learning. In a pretty definite sense, educators are a part of the methods in the promotion of learning.

In teaching agriculture, there are many opportunities for using a variety of methods. Agriculture is taught to individuals, groups, and mass audiences. The setting for the learning may be on the farm, in the agri-business, in the home, in the office, or in a class or meeting room. Both men and women from rural and urban centers who need our educational services are included in the program.

> Young people and adults can learn in different ways—from classroom instruction and from less formally structured environments—particularly in connection with their jobs. If the objective of the instruction is to master a particular technique, it is likely that they can make greater progress in a real live situation than in the simulated environment of a classroom, especially in a school that has no organic relation to the work setting (26, p. 21).

SOME POSSIBLE METHODS

Many methods are used in teaching agriculture. A teacher seldom resorts to one method in any given meeting; for example, he may use part of the period for lecture combined with audio-visual materials, and then use the rest of the period for discussion. In another situation he could conduct a field trip, make use of a resource person, and have a student give a demonstration. The following is a list of some of the methods used, classified according to individual, group, or mass basis, which illustrates the extensiveness of choices available:

Individual	*Group*	*Mass*
Conference	Problem	Bulletin
Letter	Lecture	Radio
Telephone call	Discussion	Television
Farm supervision	Panel	Circular letter
Teaching machine	Symposium	Newspaper
Project	Tour	Magazine
	Demonstration	Exhibit
	Film	
	Resource person	
	Role playing	

Many of the methods listed are used successfully, but effective teachers are never completely satisfied with the methods they use. Likewise, it can be concluded that there is no one "best" method or methods.

FACTORS AFFECTING METHODS

Each teacher needs to consider a number of factors in arriving at the best method(s) to use in any given situation. There is no formula that can be followed. The factors listed here should be considered in making the decisions.

Nature of Subject Matter Taught and Objectives To Be Attained. Teachers of agriculture know that there is much variation in the subject matter included in workshops, seminars, or courses of agriculture for adults. If the purpose of the teaching is to develop some skill in welding, the teacher would probably be able to demonstrate various welding techniques and then provide practice for the students. When the development of skills or abilities is desired, it is necessary to provide opportunities for the students to practice the correct behavior. This demonstration and practice method would not be used in developing interest and an understanding of the kind of agricultural policy program that would be good for farmers. In such a subject matter area the use of discussion, lecture, and resource persons are more appropriate methods.

Number of Students Involved. If the group of adults to be taught includes one hundred or more people, it appears that it would be logical to use a lecture, movie or slide presentation of some nature. A large group is not as suitable for discussion as a group including ten to twenty persons. It is quite obvious that if one individual is to be taught, such methods as conference or demonstration technique would be used.

Time Available. Time is always somewhat limited and has its effect upon the kind of methods used. If there is ample time, the teacher is more likely to use the discussion method and other techniques that encourage maximum student participation. If, on the other hand, time is quite limited, the teacher will probably lecture in order to share as many of the facts as he can.

Facilities and Materials Available. The kind and extent of physical facilities and instructional materials available—including community resources—affect the methods of teaching that can and should be used. If the teacher of agriculture has a large, well-equipped shop available, he may want to have several tractors in the shop for use in developing the students' skills of maintaining tractors properly. In another situation, a state extension specialist may be used in helping a group to make correct decisions in rural zoning. Teachers should not fail to consider the availability of farms, greenhouses, and other agri-businesses within the community to be used as laboratories for the study of some specialized developments.

Interests and Abilities of Teacher. Most teachers have personal preferences and more security in using selected methods. Other things being equal, the teacher should use the method that he likes or uses best. This does not mean that he should not be sensitive to other developments that supplement or improve upon the methods that he uses frequently. Much new "hardware" and instructional materials have been developed to aid in the learning process.

Effectiveness of Method. Generally, selection of a method or methods is dependent upon what research has found to be the best way of teaching. Thus, it is somewhat paradoxical to find that "little has been done to develop teaching methods on basis of scientific knowledge of learning" (27, p. 465). Even though research does not tell us what is best, all teachers should evaluate methods used in terms of the objectives to be accomplished in the situation at hand.

OTHER PHILOSOPHICAL AND PSYCHOLOGICAL FACTORS.

Many teachers do not realize that there are some philosophical and psychological factors that influence their choice of teaching methods; they operate within their philosophy or beliefs concerning such things as meaning of democracy, nature of man, understanding of the learning process, and purposes of adult education programs in agriculture. As Lammel points out (15, p. 90):

> The methods used by a teacher at any one time arise from his values, his conception of individuals, his perceptions of how learning occurs, and his knowledge of what will be rewarded in the area under consideration. In light of modern educational perspectives, no group of teachers can avoid a comprehensive thoroughgoing analysis of such matters.

Meaning of Democracy. Most teachers in the United States believe that schools should be operated in accordance with the values to which they subscribe in a democracy. They do not like autocratic administration. They believe that, in the curriculum, consideration should be given to the various forms of government, with emphasis upon the advantages of democracy as our social ideal. We sometimes forget that the most vivid lesson taught concerning democracy is the way we behave in a classroom or group situation, and/or the way the organization of the students is conducted.

There is probably a great deal of difference in our interpretations of the meaning of democracy. This is understandable because democracy is in a continuous state of modification. We are constantly trying to interpret ways and means of carrying out our basic beliefs.

There is general acceptance of the idea that man is of utmost importance and that regardless of race, color or creed opportunities should be provided for his optimal development. This is really the supreme test of any new procedure that we enact in trying to regulate society. Smith states (21, p. 36): "In a democracy the individual is of ultimate worth . . . Destroy everything but conserve the dignity of the individual and he will work to regain his losses. Destroy the dignity of the individual, and all else will eventually be lost."

Democracy places great faith in the intelligence of the common man. You and I and everyone else in a democracy should have a voice in making decisions and in arriving at answers to our problems. We believe in solving our problems through the method of intelligence, which includes securing and evaluating facts rather than turning to bias, prejudice or tradition. We likewise believe in sharing and working together for the good of all.

If one accepts these basic beliefs, then the teaching program should be planned and conducted with consideration for the interests and needs of all those enrolled:

> Our devotion to equality does not ignore the fact that individuals differ greatly in their talents and motivations. It simply asserts that each should be enabled to develop to the full in his own style and to his own limit. Each is worthy of respect as a human being. This means that there must be diverse programs (methods) within the educational system to take care of the diversity of individuals; and that each of these programs (methods) should be accorded respect and stature (10, p. 81).

The Nature of Man. A teacher's concept or understanding of the nature of man will have some effect upon the teaching methods he uses. Most people agree that man is an unusual organism capable of using rational powers to achieve personal goals and fulfill his obligations to society. Each person is unique. "Perhaps nothing is as clearly established in psychological research as the fact of individual differences. Each learner has his own unique pattern of capacities" (23, p. 53). Even though each person is unique, he is challenged by developmental tasks which are in the same general pattern and sequence as those faced by others. In fact, man is a social being. He needs to feel that he is accepted in a group and that he has the capacity and ability to help others in solving the problems of society.

Another of man's characteristics is that he is dynamic and goal-seeking. He is actively involved in trying to structure his environment so that life can be more acceptable to him and his close associates. "Thus, man seeks not merely the maintenance of a self but the development of an adequate self—a self capable of dealing effectively and efficiently with the exigencies of life, both now and in the future" (2, p. 45).

It must be recognized that man's behavior is organismic, meaning that a person is a complete unit comprised of mental, physical and emotional aspects. When we teach, we are teaching the "dynamic whole"—the entire person—rather than developing him only mentally.

The above statements concerning the nature of man imply that we need to give emphasis to the development of thinking, encourage the establishment of worthwhile goals and provide the kind of social setting that is acceptable.

Purposes of Schools and Educational Programs. There are many purposes of schools and educational programs, including the development of interests, appreciations and abilities contributing to effective citizenship. Most programs give support to efforts that aid the individual to live harmoniously and efficiently. "The American people have traditionally regarded education as a means for improving themselves and their society. Whenever an objective has been judged desirable for the individual or society, it has tended to be accepted as a valid concern of the school" (8, p. 1). Frymier points out (9, p. 7):

> The purposes of education imply direction. They assume that behavioral change is both an appropriate and a legitimate goal. Schools seek to effect such changes through communication; through persuasion rather than coercion; through influence rather than physical force. In a democracy schools are committed to specific ends, but the ends are always predicted upon the dignity of man and the power of ideas. And the means must never violate the ends pursued.

Within this broad framework we have schools and programs providing a general education that is good for all. Some schools, however, emphasize preparation for the world of work more than others.

In programs where abilities and skills are emphasized, it is necessary to individualize the instruction to the extent that each person is able to participate in accordance with his capacities. This involves close association and participation experience in real situations. The so-called academic programs usually resort to methods that are primarily for the purpose of changing attitudes and understandings. Such diversity in programs exists at the adult education level as well as in the more formal programs of elementary and high school education. For the most part adult education in agriculture is concerned with those methods that involve problem solving and doing to learn.

Some Principles for Promoting Effective Learning

There are many principles of learning. However, our discussion is limited to a consideration of only five of the most important in terms of providing conditions for effective learning on the part of adult students in agricultural programs.

WE LEARN BEST WHEN THERE IS INTEREST

It is well established that we learn more effectively if we have interest or an optimum level of motivation.

> All educational institutions concern themselves over the problem of what impels their students to learn. Learning is work, sometimes very hard work, and it looks easy only when it is either not going on at all and the students are merely being entertained, or when motivation is so high that the work involved in the learning task becomes enjoyable as well as arduous (17, p. 38).

Interest is a personal, dynamic quality that motivates us to spend more time and effort on that which we find desirable. If interest does not exist in that which we are attempting to teach, then it must be developed by relating our teaching to the interests of the learners. Among other things, adults in agriculture are concerned about making more money, having higher production and providing for family needs including a higher standard of living.

In adult education we have two basic problems with reference to motivation and interest. First of all if we are to have a program of adult education, we must have people to make a start in attending our meetings; and once they start, they must be so motivated that they will return for successive meetings. All of us recognize that there is keen competition for time among the many social, recreational and community activities. Such activities as watching television and reading magazines, which in many cases are interesting and good, must be looked upon as second best in comparison with participation in the adult program. It must be remembered that adult education is voluntary; if people do not choose to participate, we do not have a program.

We might assume that we as teachers in adult agriculture know what the program should be and that we have the necessary professional competence to plan it, announce it to the community and carry it through to completion. We may assume further that the program will be so good that a number of the people we want to serve will come to the meetings without any further effort on our part to motivate the prospective enrollees. Such assumptions and plans have led to many disappointments in the number of people participating.

Most of us have found that we must approach the planning of adult education programs knowing that interest needs to be developed. Even though there is no one best way that this can be done, we recognize that there should be some communication with our prospects. Nothing seems to be more important than personal contacts and consultation. We need to show an interest in the people who are to be enrolled and to secure from them an indication of their interests. This is a worthy teaching objective in itself. Making personal contacts is time consuming, and teachers need assistance in such an effort if a sizable group is to be served. This assistance usually comes from a planning or advisory committee, which is discussed in Chapter 5.

The motivation that is necessary for students to continue in adult education programs is provided by effective teaching. Students who are having their needs met and who enjoy the total meeting experience will return for successive sessions. There is no formula for motivation. The following procedures, however, seem to be workable in most adult learning situations:

1. Base the teaching upon the needs of the students. What is taught should be meaningful to those who are participating. It should help them reach goals that they look upon as being more satisfactory than the present attainment.
2. Base the teaching upon thinking and understanding, rather than memorizing facts. Emphasize the use of facts in solving problems.
3. Make use of the natural drives and urges that exist in most people. Appeal to ownership, desire for recognition, curiosity and suspense, activity, competition and group participation.
4. Use illustrations that provide connections with that which is interesting, including persons and happenings.
5. Make use of appropriate audio-visual aids in teaching. This may necessitate taking the individual or group to the farm or to the business.

6. Carry the teaching to the doing level. Be sure that the students are able to apply that which has been taught. This will in many cases necessitate individual follow-up on the farm or in the agri-business situation.
7. Provide favorable physical facilities wherever the learning situation takes place. An adult student cannot become interested in that which is being taught if he is uncomfortable and more concerned about the heat and the light, for example, than the problem being discussed.
8. Teach enthusiastically. Be concerned with each individual and the problems that are being discussed.

Other devices that contribute to motivation include starting and stopping meetings on time, and having refreshments and/or other activities that are of interest. On many occasions a young farmer has been attracted to a program because after the meeting there was opportunity to play basketball or volleyball. The refreshments and/or other social experiences before and after the meeting may be the real interest of some in starting a program. This should not be looked upon as being undesirable. The "sugar coating" that is provided is a part of the total value of a program.

Another motivating force for adult education in agriculture is the organization or "learning group" of the students. This resource is effective in supplementing the instruction and guidance of the teacher. If a group decides to support some educational activity or experience, there is a pressure that each student seemingly places on his colleagues to co-operate in the venture. For example, a young or adult farmer group may become organized and develop activities such as a 150 Bushel Corn Club. As a group, they develop their goals and rules or policies to be followed. They do all they can as a group and as individuals in supporting the kind of educational program that will provide them the ways and means to accomplish their goals. As Miller states (17, p. 41):

> But we have available to us another resource which is often over-looked or used without much skill: the forces of the learning group itself arising out of its attractiveness for the members of the group, and the quality of their interaction. There is no question at all of the power of the group to reduce individual resistances to change if properly mobilized. . . .

All adult programs should provide a worthwhile social experience. A person will continue to participate in a meeting only if he is accepted socially, and there is something in it that he likes and that meets some of his goals. "The significance of social acceptance relations in the learning activity lies in the fact that each learner needs approval. This need is satisfied through the feeling that he is important as a person and that he is accepted as a person" (12, p. 280).

We Learn Best When Needs Are Being Met

Generally, teaching is effective when it aids the student to progress in the accomplishment of goals for which he is striving.

> To be most meaningful, learning should be aimed at achieving the goals of the learner and in helping him move toward their attainment. Nothing sustains the adult in learning more than a feeling of getting somewhere. And a feeling of getting somewhere is a compound of perceived early achievement which is moving in the direction of the learner's objectives. In other words meaning is in part a function of the relation of learning to the learner's goals (12, p. 170).

Meaningful instruction meets needs and makes sense to the student. Instructional information is for the purpose of solving problems rather than only adding to the storehouse of knowledge. Dr. Alfred N. Whitehead, Philosopher of Education, undoubtedly was thinking in these terms when he made the statement (29, p. 1): "A merely well informed man is the most useless bore on God's earth." It isn't what you know, but how you use what you know.

If needs are to be met, the student must see that his present practices and behavior are unsatisfactory and that there are better ways available to him. He must have arrived at a decision that change is not only possible but necessary, and that he is willing to devote time and effort to bringing about a change. In addition to being dissatisfied with his present status, the student needs to know what the better behavior is like and why it is necessary or desirable for him to make the change. Unless he can see this rather clearly, he is not likely to support a program of adult education. Perhaps there is nothing more important in adult education than developing this attitude of wanting to do something. Once the student wants to accomplish a certain objective, he is likely to pursue an activity for its accomplishment regardless of the alternatives provided by the teacher.

The teacher must plan to help students perceive and analyze their needs by taking inventory of their practices, knowledge and abilities. It is also necessary to know each student and his individual performance as a farmer or agriculturalist. It means making observations, raising questions and getting answers concerning goals, practices and outcomes. It means sitting with a farmer, for example, analyzing his business records and attempting to identify any weaknesses that they may reveal. Agriculture teachers continually encourage farmers to keep records and to analyze them in terms of their present and future operational efficiency. This type of record keeping is basic and important. In many cases it is a long-term program which necessitates a good understanding and personal relationship between the teacher and students.

Some agriculturalists, like other people, do not readily admit some of their problems openly. They sometimes feel that they are the only ones who have such problems. Teachers need to assure them that the recognition of problems and dissatisfactions does not indicate a weakness but a strength. All people have problems, and those who accomplish most are those who are willing to face their problems and do something about them.

> Instruction must be exceedingly sensitive and flexible in this kind of situation if it is to succeed in helping students to recognize their own inadequacy. Some groups are willing to do so directly, but others need to approach much more indirectly the analysis of what constitutes adequacy in their particular context. Case studies from some not-too-closely connected field might be a better start for some groups; or, if the particular instructor happens to be skilled enough, he can start by accepting the group's definition of their problem as belonging primarily to "all those other people" and help them to arrive at some insight into their own part of the problem (17, p. 43).

The problems of an individual or a group lie between where they are now and where they should be. Therefore, in addition to identifying their current situation, students need to see what is ideal, where should they be, what are some reasonable goals and what others in similar situations have accomplished. There is no one best way to become knowledgeable about such outcomes. Generally it is well for teachers to share with individuals and groups the latest research findings and point out yields, good practices and other outcomes that are feasible. Students should attend experiment station field days, participate in tours of selected farms or observe demonstrations to see what can be accomplished through the application of better practices. Somehow the students need to raise their sights to improving their situations.

Once a person is aware of better behavior, practices and outcomes, he will be interested in knowing what is involved in putting them to use. What does it cost in terms of money, time, study and effort? What might I expect? Is the effort worthwhile? Would this be the best use of my time, money and labor? The student will be searching for answers to such questions. The providing and analyzing of this kind of information is the reason for programs of agricultural education. If personnel, such as farmers, understood what they needed to know about the most recent research and developments, teachers of agriculture would not be necessary. In a sense, the entire agricultural education program is a "middle-man's operation." It helps interpret the new and better situation so that interested persons can make improvements.

The acceptance of the concept of meeting needs necessitates planning programs where theory is integrated with practice. It means that the teacher must understand the latest research and practices that should be implemented and know the situations wherever the learning is applied. Problems exist in the gap between the research and the current situation. This is the challenge and the work of the teacher of agriculture.

Many successful adult education teaching and learning experiences are based primarily upon meeting the student's needs. For example, an adult farmer meeting was centered primarily on the development of understanding and abilities to use a balanced ration in feeding dairy cattle. This meeting was part of a series of twenty meetings on dairy management which included studies that were identified and planned by the adult farmers and the teacher as most appropriate. (Incidentally, the planning of the curriculum is very important in determining the extent to which needs are met.)

In preparing for the meeting, which was to emphasize the grain and supplement part of the ration, the teacher felt that it was best to use a specific situation of a member of the group. He visited Frank Schirm's farm about two weeks in advance of the meeting. In discussing Frank's feeding program, the teacher discovered that it was fairly good; however, it did not contain enough digestible protein. Frank agreed to let the teacher use his problem as the basis for class discussion.

Very early in the meeting the teacher placed Frank's ration on the chalkboard and asked the members of the group what they thought of it. The members raised various questions about Frank's returns in milk production and the kinds of grain and other feeds that were available. In reply to a question about the kind of hay being fed, the teacher displayed a bundle of Frank's hay. Immediately, someone commented that the quality of the hay affects the kind and amount of grain and supplement that should be fed. The standard amounts of digestible protein needed in the grain ration with hay of different qualities were then presented. (Before the meeting the teacher had placed this needed information on a rolling chalkboard that was in front of the room. After seeing these standards, the class began to analyze Frank's ration. Several of the farmers who were more keen in mathematical calculations determined the percentage of digestible protein in the ration. The class found, as the teacher had found previously, that more protein was necessary.)

The next subject discussed was the possible ways of getting the needed amount of protein. Factors such as suitability, costs, amount of protein and accessibility were considered. Before the discussion was concluded, two logical answers to the problem were agreed upon.

Frank had the advantage of others helping him to think through his problem, and the other dairymen received new information which helped them to solve their problems. Before adjourning, three or four others who were feeding different kinds of hay and silage indicated that they would like to have some analysis of their feeding programs. This was done at the next meeting, but before concluding, some generalizations and emphasis upon principles of feeding dairy cattle were made.

These meetings were interesting, thought-provoking and practical. They helped the members of the group to see what they were doing wrong and how to improve their practices. It should be mentioned that this was a mature group that knew each other well; they felt free to share experiences.

We Learn Best When Thinking Is Stimulated

As implied in the foregoing illustration, individuals who will make decisions and learn most from the adult education experience need to think through their problems themselves if their needs are to be met. Most teaching is effective to the extent that it stimulates thinking. Dr. Boyd Bode, an educational philosopher at The Ohio State University, frequently said, "Thinking is basic, it is the educational kingdom of heaven—if we have it, all other things will be added."

Simply stated, thinking is the recalling and using of facts. Reflective thinking occurs only as a result of a challenge. It does not occur in a vacuum, but rather in terms of learnings and experiences. Therefore, teaching and learning must aid the student in relating and integrating the new facts with the old.

Understanding is very closely related to thinking. Stewart's concept of understanding is that it involves the meaning of terms and seeing cause and effect relationships (22, p. 107). He points out that the more relationships one is able to see, the more he understands. For example, a farmer may understand that the use of fertilizer will increase his yield of corn. He may also know that the fertilizer contains nitrogen, phosphorus and potash. He may also realize that if he applies the fertilizer too close to the kernels of corn, it will damage the plant, but he does not have as much understanding of the process as an agronomist who has studied more chemistry and plant physiology. With his additional background, the agronomist would know how the various nutrients help to promote plant growth, and he would understand the need for additonal plant nutrients. Therefore, when someone says he understands, it means that he has *some* understanding, but not a *complete* understanding.

Basically, thinking is a problem-solving process. It includes encountering a difficulty or frustration, seeing possibilities or alternative solutions, securing facts, evaluating them, arriving at a tentative decision, trying it out and then finally adopting it. John Dewey (7, pp. 152-163) called this process "a complete act of thought." He advised that an educator should teach to develop such thinking.

Adults can be taught more effectively by using problem-solving procedures more extensively, both in group and in individual situations. Problem-solving teaching is interesting to the learner, it relates theory to practice, and it will more likely be carried to the "doing level" than other methods of teaching. Here again it must be emphasized that one cannot use the problem-solving technique effectively unless he knows the situation that needs to be improved. Hammonds and Lamar indicate the following reasons for using problem-solving in teaching (11, pp. 80-81):

1. When problems are used in teaching, the learner has an idea of what he is to do. He knows the problem to be solved, what is to be considered or learned, and perhaps the kind of information needed. Most students learn very laboriously or not at all unless they know these things. The statements just made do not mean that in problem solving the learners follow directions.
2. The problem acts as a selective agency in gathering pertinent facts and organizing them. It tends to favor the perception of facts appropriate to the solution of the problem.
3. In problem solving, significant facts are taught in useful association and thus will be more likely to be used when the need for them arises. These facts take on a meaning they would not have otherwise; they become items of experience. One's knowledge of a fact is its residual effect on him, of what he has done with the fact, the use he has made of it. In problem solving, the teacher and the students get an idea of relative values of the fact or thing being considered.
4. Problem solving places emphasis on use of information rather than on memorization of it.
5. Where problem solving is used, there is creativity in teaching. Each step or phase of problem solving is a creative activity.
6. Solving a problem calls for the use of old learning in new ways. Thus there is desirable practice of, or experience with, what is to be learned or retained.
7. Problem solving in teaching contributes to developing in students a habit of evaluation and using data intelligently.
8. Problem solving helps the students learn to keep their thinking in the field of the problem.

9. As problem solving calls for pause and the weighing of alternatives, it contributes to the development of open-mindedness. The only way to teach a student to pause, weigh, and examine is to have him do these things.
10. Use of problem solving in teaching should contribute to developing the ability to discover problems. Problems do not arise; they are discovered.
11. Problem solving lends itself to learner participation. The teacher has the opportunity to direct or guide the learner's activity—which is a large part of teaching.
12. Decisions made in group problem solving are more likely to be carried out than those made in some other situation, especially if it is known that others in the group have decided to act.
13. The many variations in kinds of problems and methods of solving them give rich opportunity for flexibility in teaching.
14. Using problem solving in teaching increases the ability to solve problems, which is intelligence.

Other aids may be used to develop thinking, understanding and interest, including a variety of instructional materials. Most of us can better understand those things of a concrete or objective nature than those of an abstract nature. A field trip or a demonstration in many cases will develop more interest and understanding than a lecture. Audiovisual aids, if properly planned and used, will facilitate more learning. One teacher illustrated this point by describing a newly developed apple. He then drew the apple on the chalkboard with white chalk, emphasizing its shape and size. His next step was to use colored chalk to typify the coloring. To make the concept still more vivid and meaningful, he reached below his desk and showed the group a real apple, and then he passed samples to all of the individuals so they could see it, smell it and taste it. There was no question that much more understanding was developed as a result of employing all the students' senses in the teaching process.

Currently many instructional media are being developed. These media aid learning both in groups and in individual situations. Taylor and Christensen point out (24, pp. 17-18), ". . . since the instructor's role is to help the student help himself, attention is shifting from teaching to learning with the accompanying emphasis on flexible scheduling, flexible learning space, and self-instructional devices." They indicate that private industry is providing many electronic devices to aid the teacher. Recent mergers of a number of electronic firms have brought into being an increased capacity for developing learning systems. Adult teachers need to keep alert for such things as tapes, film loops and packaged learning devices.

> Instructional packages, if carefully developed, could be used to achieve vocational teaching objectives. . . . A carefully designed package will exploit the instructional advantages of several media. Suppose, for instance, the instructor wants to teach the procedures involved in making quality concrete. He might introduce the unit with the test and supplement it with a company-sponsored film. The instructional package, however, might also include cutaways of actual concrete blocks of various qualities, the model of a slump tester or sediment jar used in the process, slides to teach aspects of the operation that require longer scrutiny, single-looped films to show concepts or operation where motion is important—or any number of other combinations, but in each medium used only for the unique purpose to which the contribution is maximizing the learning process (24, p. 19).

Programmed-learning devices are usually made available for individual use by county agents and teachers of agriculture as a part of their libraries and laboratories. In addition, closed-circuit television programs can be planned and disseminated throughout various areas as a means of making specialists available in local programs. These aids will help students to solve their problems, and help the teacher of agriculture keep up to date with current information and practices.

Most learning situations should include sequential instruction and instructional materials on a level of understanding that the student can grasp and use in sequence. "Material that is well organized and meaningful to the learner will be most efficiently and effectively learned, retained and transferred" (18, p. 367).

Instructional materials should integrate theory and practice. Illustrations of practical application should be used extensively in helping to make the transition from theory to practice, and particularly in developing understanding of principles and transferring learning to new situations. John Dewey stated: "A theory apart from an experience cannot be definitely grasped even as a theory" (7, pp. 333-345).

> We can increase transfer by practicing our new learning in varied context by noting many illustrations of a generalization in addition to those in the textbook. Most students when asked to illustrate oxidation will mention only the rusting of iron. If a newly learned term or principle is to be widely transferred, become rich with association, many applications must be noted and practiced. We use it or we lose it.
>
> We can increase transfer by generalizing or intellectualizing our experiences by developing concepts, by searching for an emerging principle. The habit of seeing relationships and unity in apparent

diversity can be widely transferred. Learning by discovery gives practice in building insight (6, p. 2).

With the great rapidity of change in agriculture, it is highly important to know *why* as well as *how*. Problem-solving methods—where questions are raised by the group, where information is analyzed with them rather than for them and where facts are evaluated and answers arrived at as a group—will develop the kind of understanding that is necessary for the adults to be self-directing learners. The approach of starting with problems, analyzing a number of them and then arriving at the principles or generalizations seems to be much more interesting and effective than to "dish out" the facts and information or principles with the hope that the students will understand.

Agricultural education needs to develop more materials primarily for adult learning. It has been somewhat handicapped in this regard, because most of the materials have been prepared for high school or college students. In the development of materials, more attention needs to be given to the specific educational objectives to be attained. This holds true for all materials such as bulletins, circular letters, supplementary guides and forms that are used in both individual and group teaching situations.

We Learn Best When Active Participation Is Provided

Learning is a self-active process. Each person learns for himself to the extent to which he participates. The teacher cannot learn for the student, but rather sets the stage so that the student will learn more efficiently with the procedures and materials provided. This concept is not unique to teachers of vocational education or to personnel in agricultural education. "For many years in American educational literature the concept learning by doing has been almost a universally accepted article of faith. Teachers and prospective teachers continually are reminded that learning is an active affair and we learn by doing" (19, p. 338).

A class or meeting is usually better if it provides an opportunity to participate. Even though they know this to be true, too often teachers have monopolized the class, and have done most of the thinking and studying rather than challenging the student to do it. Teachers are inclined to explain too much and resort to the old pattern of imparting knowledge.

Learning through activity means more than physical involvement. "Activity includes all that one does . . . his overt actions, his thinking, his feeling, his perceiving, his imagining, his comprehending, his relationship seeing . . ." (11, pp. 177-78).

Agriculture teachers have an advantage over most teachers because they can see evidence of the learning that has taken place in terms of changes made in performance. For example, the farm, the greenhouse and the agri-business are laboratories where the learning should be applied. This does not mean that the growing of new varieties of mums is the only indication of learning. The florist had some interests and understandings before he actually selected and grew the new varieties. Learning in agricultural education is a combination of cognitive, effective, and psychomotor skills in terms of behaviors. In many cases the teacher of agriculture is interested in promoting all of them—thinking, feeling and doing.

In vocational agriculture and extension, educators have commonly used the project method of teaching in which teachers help their students to select and conduct project experiences as a means of furthering learning and improvement. Perhaps this method is best understood in terms of its use with 4-H Club members and high school students of vocational agriculture. Generally, teachers insist that projects or occupational experience programs be closely related to the class and group instruction, that the students have managerial and financial involvement, that their programs be increased in size, scope and complexity in accordance with their capacity, and that they keep and analyze records of their experience. This method has proven to be desirable for integrating theory and practice. The very practice, in most cases, adds interest for still more learning.

Adult education students likewise have opportunities for laboratory experience. Agronomic demonstrations, for example, have developed much interest in corn production programs. Farmers electing to engage in such demonstrations have been anxious to identify new and better practices, apply them, observe their own crops as well as those of others, evaluate results through carefully conducted yield checks, and then analyze their findings. Some groups have found that the initial interest was based upon getting high yields and competing with other members of the group. In most cases, however, the interest shifted to include a consideration of the economic factors involved. In addition to the selection of a single enterprise on which to concentrate, many teachers and their farmer students look upon the entire farm business as the project through which they continue to learn as well as earn. This is particularly true in farm business and planning programs.

The above illustrations imply that the teacher of agriculture should relate his off-farm instruction to on-farm practice and development. Individual instruction, which is discussed in some detail in Chapter 7, is very important in promoting the learning process through participation. Other methods that emphasize activity on the part of the learner are field trips, discussions, student reports, and the work of committees. The interesting aspect of this concept of education is that the adult teacher who involves his students will promote more learning and at the same time secure assistance which he needs in planning and conducting the program.

We Learn Best When Correct Behavior Is Reinforced

In all effective learning there must be some reinforcement or reward of the correct or desirable behavior. Learning succeeds best in a climate of success. This does not mean that one must always succeed to learn. But unless success is generally associated with the new behavior to be practiced, the learning program may be discontinued—particularly in adult education.

Much has been said and done in recent years concerning programmed instruction. The effectiveness of this learning device has been largely due to the consistent use of specific learning objectives, sequenced material that is appropriate for the objectives and immediate reinforcement. Miller points out (17, p. 205):

> The key is in the behaviorous principle that any live organism will tend to repeat a particular behavior if that behavior has been followed by a reward, that is, if it has been reinforced. The essential characteristic of programmed materials is that they arrange concepts or relationships which are to be learned in a series of short steps which require the learner to make appropriate responses. The material itself provides immediate reinforcement by telling the learner that he is correct. Stimulus, student response, and immediate feedback are the core of the method.

There is no one best way to provide rewards or the reinforcement that the learner needs. However, the teacher should make such reassuring statements as: "That's very good;" "I agree;" "That's what we have found;" "This matches what they are using at the demonstration center." These kinds of statements as well as nods of approval and general acceptance of peers in group situations provide part of the motivation students need. Students also need to secure an evaluation of what they are doing on the farm, in the home or in the agri-business where the learning is being applied and continued. The student must be made knowledgeable about mistakes, wrong practices and omissions that should be corrected. One of the real challenges for the teacher is to correct mistakes in such a tactful manner that the learner will not be discouraged but will be appreciative and stimulated to continue to make progress.

In most educational programs, some repetition is desirable. It can be achieved in meetings with adult students by making applications to different situations, by making brief summary statements of the previous meeting and by giving recognition to students who have initiated a practice that has been emphasized. It may sometimes be necessary to repeat certain sections of an audio-visual presentation because the students may not have learned as much as the teacher assumed, as found by Cook in some research that he summarized (3, p. 99):

> The concensus of the findings is that audiences can usually profit from a greater number of repetitions than they are generally given. The person who creates an audiovisual demonstration is usually so familiar with the material that it may be difficult for him to believe that people cannot learn it in one trial.

An environment for effective learning should be free from emotional distress. Problems and situations may occur previous to an adult meeting or conference that are not conducive to learning. Illness in the family, financial reverses, disagreements, failures or fear of failure and lack of social acceptance are occurrences that inhibit the effectiveness of learning. Teachers, therefore, need to be alert to happenings and sensitive to reactions coming from students who are not responding according to expectations.

Practice does not necessarily make perfect. Students can practice a behavior that is wrong as well as that which is right. A welder does not become better by practicing incorrect techniques. Teachers must make sure that the students know the correct behavior and then must provide what some educators refer to as "sequential practice":

> The learner should have opportunity for a good deal of sequential practice of the desired behavior. In sequential practice each subsequent practice goes more broadly or more deeply than the one before. Usually, each additional practice or experience should have some new elements in it, if it is to result in the most effective learning. Concepts and principles need to be brought in again and again in new and varying situations. The same thing holds for developing skills or other abilities and for developing appreciations. High level learning and high level teaching call for sequential practice (11, p. 197).

This implies that adult programs should proceed from the more simple to the more complex, in a series of meetings based upon a series of courses that are developed on a long-term basis. It implies, too, that the teacher will aid the student in setting successively higher standards of performance and assist him in evaluating such performance and making plans accordingly. The teacher who encourages students to take charge of their own learning is successful.

REFERENCES

1. Bruner, Jerome, *The Processes of Education.* Cambridge, Mass.: Harvard University Press, 1961.
2. Combs, Arthur W., and Donald Snygg, *Individual Behavior.* New York: Harper & Row, Publishers, 1959.
3. Cook, J. O., *Research Principles and Practices in Visual Communication.* East Lansing: Michigan State University, National Project in Agricultural Communications, 1960.
4. Dale, Edgar. *Can You Give the Public What It Wants?* New York: Cowles Education Corp., 1968.
5. _______ "Learning to Learn," *The Newsletter.* Columbus: The Ohio State University (May, 1963), pp. 1-2.
6. _______ "Principles of Learning," *The Newsletter.* Columbus: The Ohio State University (January, 1964), p. 2.
7. Dewey, John, *Democracy and Education.* New York: The MacMillan Company, 1961.
8. Educational Policies Commission, *The Central Purpose of American Education.* Washington, D. C.: NEA, 1961.
9. Frymier, Jack R., *The Nature of Educational Method.* Columbus: Charles E. Merrill Publishing Co., 1965.
10. Gardner, John W., *Goals for Americans: Report of the President's National Commission on Goals.* New York: Columbia University Press, 1960.
11. Hammonds, Carsie and Carl Lamar, *Teaching Vocations.* Danville, Ill.: The Interstate Printers & Publishers, Inc., 1968.
12. Jensen, Gale A. A. Liveright, and Wilbur Hallenbeck, *Adult Education.* Chicago: Adult Education Association of the U.S.A., 1964.
13. Krebs, Alfred H., *For More Effective Teaching.* Danville, Ill.: The Interstate Printers & Publishers, Inc., 1967.
14. Kuhlen, Raymond G., *Psychological Backgrounds of Adult Education.* Syracuse University: Center for the Study of Liberal Education for Adults, 1963.

15. Lammel, Rose "General Versus Special Methods in Teaching the Disciplines," *Theory Into Practice*. Columbus: The Ohio State University (April, 1966), p. 90.
16. Lorge, Irving, *et al., Adult Education Theory and Method: Psychology of Adults*. Chicago: Adult Education Association of the U.S.A., 1963.
17. Miller, Harry L., *Teaching and Learning in Adult Education*. New York: The MacMillan Company, 1966.
18. Pressey, Sidney L., Francis P. Robinson and John E. Horrocks, *Psychology in Education*. New York: Harper and Brothers, 1959.
19. Rich, John M., "Learning by Doing—A Reappraisal," *High School Journal* (May, 1962).
20. Schroeder, Wayne L., "Adults Can and Must Learn," *Journal of Cooperative Extension*, Madison: University of Wisconsin (Winter, 1966), p. 207.
21. Smith, William S., *Group Problem Solving Through Discussion*. Auburn, Ala.: Auburn University, 1963.
22. Stewart, W. F., *Methods of Good Teaching*. Columbus, Ohio, 1950.
23. Stratemeyer, Florence B., *et al., Developing a Curriculum for Modern Teaching*. New York: Teachers College, Columbia University, 1963.
24. Taylor, Robert E., and Virgil E. Christensen, "Instructional Media in Vocational Education," *American Vocational Journal* (January, 1967), pp. 17-18.
25. Thorndike, Edward Lee, *et al., Adult Learning*. New York: The MacMillan Company, 1928.
26. *Vocational Education*. Chicago: National Society for the Study of Education, 1965.
27. Wallen, Norman E., and Robert M. W. Travers, "An Analysis and Investigation of Teaching Methods," *Handbook of Research on Teaching*. Chicago: Rand McNally & Co., 1963.
28. Weir, Edward C., "Choice and Creativity in Teaching and Learning," *Phi Delta Kappan* (June, 1962), p. 408.
29. Whitehead, Alfred N., *The Aims of Education*. New York: The MacMillan Company, 1959.
30. Wiles, Kimball, "The Teacher Education We Need," *Theory Into Practice*. Columbus: The Ohio State University, College of Education (December, 1967), p. 26.

Chapter 4 CLIENTELE FOR ADULT EDUCATION

Adults will become favorably committed to adult education in agriculture when they perceive it as useful to them. They will participate in a well-designed program whether it is a formally organized course, a television series or a program of individual conferences. However, a program cannot be designed until a target audience is identified. This chapter is devoted to deciding whom to teach and how to ensure their continued participation in the program.

Deciding Who Shall Participate

Teachers in vocational agriculture and cooperative extension typically have perceived themselves as teachers of young or adult farmers. Others have perceived them in this same way. The question remains, are agricultural education programs for farmers only or can some other groups benefit from them? The following questions will be useful to professional educators in deciding who should be in their target audiences.

What Adult Education Group Was Served Last Year? It is important in the development of adult education programs to start at the students' present level. A primary concern is the evaluation of what has been done in the past. Educators must continue to work effectively with past participants as well as begin to serve new clientele groups.

The new teacher may continue working with previously established adult groups while carefully considering new programs. A new educator should not let himself become known as one who does not consider serving new clientele groups who can benefit from his teaching.

Is the Program To Be Designed for a Generalized Audience or One with Specialized Interests? Many teachers bring together a formal group of enrollees and then build a program of great variety to serve their diverse needs. Some build a specialized program by identifying a specific audience with members having common interests and needs. The same principle of target-audience identification can be useful in writing a news article, producing a television program or planning a tour. The personal preference of the students and the teacher is one factor to consider in answering the generalized - specialized question.

In Chapter 3 it is stated that the student will learn more if he is interested and the program is based upon his needs. The specialized course pattern provides the opportunity to more nearly fulfill this principle, since technical knowledge needs can be met best if all students need the same knowledge. In most cases agricultural educators should try to work with groups with specialized needs. However, in groups such as young farmers, where social needs may outweigh the highly specialized subject matter needs, the teacher may bring together people of similar ages who need a variety of technical knowledge.

What Method of Instruction Will Be Used? The task of developing people's interest is easier if the teacher only works with one method of instruction, because the target learners can be more readily identified. However, the teacher's preference will determine whether individual, group or mass media methods are used.

The teacher's responsibilities (i.e., teaching a full load of high school subjects) will also determine whether he can attempt to reach more than one target audience. Certainly every teacher of vocational agriculture and every extension agent will teach at least one formally organized adult education group each year. Extension agents should be able to conduct several adult classes simultaneously in the winter months. Whether teachers see clientele groups as identified audiences being taught via mass media may depend upon their organization's point of view regarding clientele groups.

What Is the Organization's Point of View on Clientele Groups?
Neither vocational agriculture nor cooperative extension claims to be able to serve all people, even though all people in society are directly or indirectly influenced by agriculture.

The teacher of vocational agriculture is restricted only to teaching ". . . vocational education in any occupation involving knowledge and skills in agricultural subjects, whether or not such occupation involves work of the farm . . ." (1, p. 43). The Vocational Education Act of 1963 and the Vocational Education Amendments of 1968 provided specifically for the teaching of adults—both in production agriculture and in related businesses. Vocational educators should consider expanding their clientele groups to adults other than just those in production agriculture.

Some practical considerations may limit the expansion of adult audiences: (1) not enough adults in the community are in a specific occupation; (2) teachers have limited professional preparation to work with a new audience; (3) time does not permit expansion beyond the present program. Although these considerations may limit expansion, teachers must consider alternatives to these problems, such as countywide programs in which one teacher takes the responsibility for working with farm machinery dealers, while another works with the nurserymen. They should also plan jointly with their county agents to serve the needs of those in agri-business.

Certainly cooperative extension personnel have had the opportunity to serve many audiences beyond those in production agriculture. The Smith-Lever Act many years ago defined their work as follows: "In order to aid in diffusing among the people of the United States useful and practical information on subjects related to agriculture and home economies. . . ." To some degree, agri-business audiences have been served by cooperative extension personnel, especially through the multi-county, area center or state specialist programs. Some recent research (3) supported the idea that programs must be specifically designed for those in the agri-business, with no attempt to involve both production clientele and service personnel in the same programs.

The cooperative extension and vocational agriculture services have both promoted work with adults in production agriculture and related industries, but they have not served all the needs of adults in agriculture. Since agri-business people are providing increased service to farmers, they need to be well educated in basic concepts and principles of the agricultural sciences—an education which extension and/or vocational education can provide.

Are There Enough Learners To Make the Method Effective? The needs of one family may be sufficient to require a program of farm business planning and analysis—certainly four or five families' needs would be ample, permitting some efficiency through small group teaching. A television program on marketing information for consumers may not be worthwhile unless it has a twenty-thousand-member viewing audience.

These examples are given to point out the need to assess the target audience's size in relation to the methods used. Most groups should have from ten to thirty members. Large groups should be taught via other methods than just group discussion or lecture.

Does the Teacher Have the Competence or Can He Arrange for Competent Persons To Do the Teaching? Neither teachers of vocational agriculture nor extension agents can be competent in every subject. They must know how to secure competent resource people, but at the same time they must be respected as teachers who are competent in some areas.

Is There Enough Time and Equipment To Serve the Potential Students? The amount of time and equipment available greatly affect the class size. The teacher must have enough time to organize, enroll, plan, teach and do intensive follow-up on the farm or in the agri-business if his class is to be effective. In a class where each student needs equipment, the teacher must have ample equipment to be able to teach effectively.

Although the preceding questions may not totally solve the problem of whom adult education should serve, they should help to focus on some target audiences. If a teacher or other educator is new to the community he can well use the general rule of starting at the students' current level, working with presently organized courses or programs if they have been successful. But he should not fail to explore new and exciting programs that he may want to initiate in the future if they are needed and the people will support them.

Prospective Adult Clientele Groups

Within the concept of agriculture, there are many clientele groups which might be served, including avocational as well as vocational groups. Since most vocational and extension educators are teaching for gainful employment, however, this book does not discuss avocational groups.

Following are some of the more promising clientele groups which might be considered as target audiences:

Association with Agricultural Production
- Commercial farmers
- Small farmers
- Part-time farmers
- Young farmers
- Older farmers
- Landlords - absentee
- Tenants
- Farm Machinery employees
- Fertilizer and lime employees
- Seed company employees
- Landscape and nursery industry employees
- Greenhouse industry employees

Associated with Marketing
- Farmers
- Truckers
- Marketing organization employees
- Processing organization employees
- Wholesalers
- Retailers

Associated with Community Development
- Planning groups
- Zoning boards
- County commissioners
- Chamber of Commerce
- Recreation Development Organizations

The groups mentioned are only a few of the many which might be served. Some specific examples of programs are: (1) Programs for the owners of forest lands which have been set up in the metropolitan areas near where the owners lived. This arrangement resulted from the recognition of a clientele group who needed education—but who lived away from the community concerned most of the year. (2) Educational programs have been held near some metropolitan areas on "How to Get Out of Farming." (3) Another progressive adult educator has directed a program for the board of directors for farmer cooperatives—a new audience, but one vital to the life of rural America.

Identifying and Enrolling Clientele Groups

Two extreme approaches to the identification of a clientele group are to (1) wait until the group asks for help or (2) wait until a supervisor decides to develop a program. Neither is characteristic of the effective adult educator. The educational innovator is the one who searches for ways to serve a new audience and the needs of agriculture while working with existing clientele groups. He always considers expanding his adult programs.

IDENTIFICATION OF POTENTIAL STUDENTS

Some of the more frequently used techniques or ways to identify audiences and secure interest include: *planning groups, personal interviews, questionnaires* and *census data*. The analysis of census data to identify possible audiences should be followed by other techniques to learn of the adults' needs.

Planning Groups. Planning groups serve two major functions—planning the program and securing participation in the program. Those who are involved in planning normally will support a program more fully; they will assist in recruiting, and they will identify others who might participate in a program.

The selection of a planning group is important. A homogeneous group of young farmers probably will not suggest working with a group of landscape nurserymen. A homogeneous planning group can be effective in identifying specific programs and learners. A planning committee of community leaders from various occupations will be able to point out a broader audience.

Personal Interview. Personal interviews with potential students might well indicate to an agricultural educator how he can serve their needs. If the interview is too time-consuming, FFA or 4-H members could interview individuals about their need for agricultural information. Personal interviews will also stimulate interest within the community.

Questionnaires. Questionnaires sent to potential students can arouse interest in educational programs. These questionnaires, developed for specific clientele, will identify their interests, and help to identify other specific groups.

Census Data. Collecting and analyzing local census data will help the teacher to understand his clientele. For example, the teacher can identify some important interest groups from data like the following, which is from a midwestern county:

Name of Enterprise	Number of Enterprises	Income
Nursery Trees, shrubs	34	$ 56,000
Flowers	23	259,000
Greenhouse vegetables	11	174,000
Christmas tree sales	21	$ 15,000
Standing timber	69	42,000
Cattle and calves	1,749	3,858,000
Hogs and pigs	965	2,243,000
Sheep and lambs	281	102,000
Dairy	1,205	11,564,000
Poultry	786	3,820,000

The most economically important enterprises in this county are dairy, beef cattle, swine and poultry. However, the number of people in greenhouse flower and vegetable production is sufficient to merit consideration for an adult education program.

ESTABLISHING ENROLLMENT PROCEDURES

Effective programs provide repetition, but also progress to new ideas or better understanding of complex ideas (5, pp. 83-86). The teacher who wants to insure the development of this progress should maintain a specific audience and have continued contact with them.

To achieve maximum learning, whether in a program of 4-H adviser education, young farmer record keeping or a beef cattle school, the teacher must assure continued participation.

Free or Fee. The question of whether or not to charge a fee cannot be answered easily. In general, the payment of even a small fee will encourage a person to participate. Although both vocational agriculture and cooperative extension serve many groups and a fee may eliminate some audiences, in most middle income communities the benefits from charging a fee far outweigh the disadvantages.

In general, fee policies for adult education should be uniform within a school; thus, in most cases a fee is charged.

Open or Closed Enrollment. Enrollment policies must be decided upon locally but must be consistent with organizational policy. All programs should be open to all qualified students. (Students who lack the basic background necessary for participating successfully in a course may be served by instituting a new course.) Adults should be given help to see which program will help meet their needs.

Preserving the Students' Interest

To maintain his audience's interest, the teacher must conduct outstanding programs. Programs that stimulate people to learn, to try out ideas and adopt profitable practices are vital to his success.

Bode recently reported the following as the most important influences on continued attendance at adult farmer classes (2, p. 43):

1. Class topics important to operation of farm business
2. Learn something at each meeting
3. Weekly reminder letters
4. Use of movies and filmstrips
5. Outside speakers
6. Scheduled tours of agricultural industries.

After developing and conducting an effective program, the teacher should attempt to stimulate further interest and participation: he can acknowledge success and stimulate further interest by awarding completion certificates and giving public recognition; he can maintain the students' involvement and support by allowing them to help evaluate the program; and he can help them to put their knowledge into action by intensive follow-up at their businesses.

REFERENCES

1. *Administration of Vocational Education—Rules and Regulations.* Washington: Department of Health, Education and Welfare, Vocational Education Bulletin No. 1, 1966.
2. Bode, John Cornelius, *Factors Which Influence Attendance in an Adult Farmer Class.* Unpublished Master's thesis, Ames, Ia.: Iowa State University, 1967.
3. Cunningham, Clarence J., *An Analysis of "In-Depth" Schools Conducted by Extension Agents in the Agricultural Industry.* Columbus: Ohio Cooperative Extension Service, 1968.
4. Sanders, H. C. (ed), *The Cooperative Extension Service.* Englewood Cliffs, N. J.: Prentice-Hall, Inc., 1966.
5. Tyler, Ralph W., *Basic Principles of Curriculum and Instruction.* Chicago: The University of Chicago Press, Thirtieth impression, 1970.

Chapter 5 PLANNING THE ADULT EDUCATION PROGRAM

Adults will participate in educational programs if they are designed to meet their need to improve their businsses or be more successful citizens. This chapter will focus on the nature of a sound educational program, the program of instruction and the process of planning a program.

Nature of a Sound Educational Program

A sound educational program for adults should meet all the following criteria:

1. It is based on *research* in agriculture.
2. It is based on the *needs* and *interests* of the learners and of society.
3. It reflects the adults' most important educational problems or *opportunities*.
4. It involves the student in *planning*.
5. It is consistent with the philosophy and *objectives* of the organization.
6. It is within the limits of *resources* available.

7. The outcomes of the program are *profitable* to individuals or communities.

These criteria serve as a guide for planning and evaluating programs.

BASED ON RESEARCH

One of the great assets of agricultural education is the abundance of research upon which to build a program: considerable basic research, as well as developmental research, is being done by experiment stations; most industries have extensive developmental research underway; and many local research and/or demonstrational programs also provide a basis for developing educational programs. The individual planning a program must carefully evaluate the research information he collects for program development, in order to use valuable information from all available sources.

Adult educators must help their students to be aware of new developments that would be profitable subjects for study. For example, unless they know about the existence of new feed additives, farmers will not request help with evaluating their use in dairy feeding. Likewise, the top management in a dairy processing plant will not necessarily request programs to help their supervisory personnel unless they believe they can be taught something helpful.

BASED ON NEEDS AND INTERESTS

A program will be successful only if it is based upon the students' *needs* and *interests*. Needs which students can readily recognize are referred to as "felt" needs. A program based upon these "felt" needs is more likely to be highly interesting to students.

It is more difficult to plan a program around "unfelt" needs, for the student must become aware of his needs before he will want to participate in a program. The task of making people aware of an idea is part of an educator's *planning* and teaching role.

Not only must the program be based upon the student's needs and interests, but it must also be based upon the needs of society and it must be consistent with trends in society. For example, a small farmer may well be interested in learning about poultry production. What does he need to know: (1) how to sustain on a small basis; (2) how to get out of farming; or (3) how to get into the poultry business on a large-scale basis? The trend in our country is toward farmers leaving the poultry business or increasing the size of their business; thus, there may not be a place for helping him remain at a subsistence level.

FOCUS ON MOST IMPORTANT CONCERNS

Since agricultural education organizations are oriented to gainful employment, students must be taught what is most important to their economic stability. For example, in a community where income from dairy is $3 million, while bees produce an income of $100,000, subjects concerning the dairy enterprise should have priority. (Of course, a community where fruit is important may find that an educational program about a bee enterprise may be important because of its impact on this other enterprise.)

Educators can ill afford to develop programs around minor enterprises, their own preferences, or the whims of a few vocal people. The most important educational problem of the community can be measured by:

1. effect on a large number of people
2. economic importance
3. community needs

These three criteria cannot be assessed alone, for a community social need, such as a park, might well be more important than the lawn problems of a home owner or even the breeding problems of dairymen. For this reason, the target audience needs to be involved in the planning.

STUDENT INVOLVEMENT IN PLANNING

Effective programs result more often when the student is involved in planning, because he can help to identify what he wants to learn. This involvement also stimulates interest on the part of our clientele and thus encourages enrollment in the program.

Considerable attention will be given later in this chapter to the involvement of learners in the planning process. The point to be made here is that nearly every adult education agency in the United States uses some system of student involvement in the planning of the program. This system includes:

1. informal discussions with individuals
2. completion of questionnaires
3. interaction and suggestions of planning committees
4. planning the entire program by committee

Frequently, when a professional designs a program without learner involvement he will experience limited participation, which may result from failure to (1) build interest in the program or (2) identify adequately the student's needs as he perceives them. The evidence of non-involvement shows up in the programs where a "curriculum director" sets up *x* number of courses, a high percentage of which are cancelled—perhaps because the lack of student involvement has caused the student to fail to recognize his "unfelt" needs.

CONSISTENT WITH THE PHILOSOPHY AND OBJECTIVES OF THE ORGANIZATION

Educators must conduct programs which reflect their organizational direction without failing to take advantage of the opportunity for innovative programs. The objectives of both vocational agriculture and the Cooperative Extension Service are sufficiently broad that educators should be able to conduct many types of programs without going outside of the present objectives of the organization.

For example, a teacher of vocational agriculture may develop programs with adults who are engaged in farming and other adult clientele groups who need knowledge in the area of agriculture. However, it would be questioned if he attempted to plan a program of adult education for bankers unless the topic were agricultural knowledge for the bankers.

Likewise, within the Cooperative Extension Service each state has identified some rather specific directions. In more than half the states it is fairly well defined that the Cooperative Extension Service shall conduct programs in agriculture, home economics and related subjects. In these states, it would be questionable to develop a program related to building codes for the commercial builders in a county. In nearly twenty states, the Land-Grant University has either closely allied or joined together the Cooperative Extension Service and the general extension service. In these instances, a county extension agent might establish a program dealing with building codes for commercial builders and utilize the resources of many colleges within the university to help conduct this program.

WITHIN THE LIMITS OF RESOURCES AVAILABLE

Available resources might be of a physical property and/or of a technical competence nature. For example, educators in vocational agriculture should be careful not to plan a program requiring the resources of a highly-specialized individual unless, of course, such an individual is readily available through specialists within the vocational agriculture service or the staff of the Cooperative Extension Service. (Of course, a teacher may have local funds to procure the help of the highly-trained specialist.)

Likewise, a staff member in the Cooperative Extension Service should not plan educational programs requiring extensive use of shop facilities. Most county extension personnel do not have ready access to welders or other physical equipment of a shop nature, and thus a program of this type normally is not taught by extension services.

The two examples imply that Cooperative Extension Service personnel and vocational agriculture personnel could and should work together in identifying programs to be taught, with each group handling those programs for which it is most suitably equipped. Such cooperation avoids needless repetitive overlap of programs when resources are limited.

OUTCOMES OF THE PROGRAM ARE PROFITABLE

Any educational program must ultimately serve a useful purpose by changing the student's behavior to bring about desired behavior. In agricultural education this purpose is the adoption of agricultural practices. Programs that do not serve this purpose might well lie outside the realm of cooperative extension and vocational agriculture services.

The Program of Instruction

The program of instruction in adult education should clearly define what is to be accomplished, how it is to be done, and how to evaluate its effectiveness. Adult education programs can be conducted without an extensive written document but an overall plan in written form will provide more direction to the teacher, the student and the community regarding the program.

THE WRITTEN PLAN

Tyler (15, p. 1) has identified four fundamental questions which must be answered in developing any curriculum or plan of instruction. These questions in an adapted form are:

1. What is the educational objective of the program?
2. What educational experiences are likely to achieve this objective?
3. How can these educational experiences be effectively organized?
4. How can the achievement of these objectives be determined?

Any written plan should contain a statement of educational objectives, an outline of learning experiences to attain these objectives, and a plan for determining the progress made toward the attainment of the objectives. These three central elements; objectives, learning experiences, and evaluation are the core to any effective educational program.

Educational Objectives. Objectives indicate the educational ends of a program to be achieved. Properly stated, educational objectives should be written in such a manner to clearly identify:

1. the *target audience*
2. the *behavior* students should be exhibiting at the end of the learning experience, and
3. the specific *topic* that will be taught

For example, two different objectives are given:

1. Commercial farmers to understand how to develop a fertility program based upon soil type and soil test recommendations.
2. Community leaders to develop an understanding of the value of a regional planning commission rather than a county planning commission.

In each of the above objectives you will note there was identified a target audience and a subject matter area about which some teaching would be done. In each instance, the behavior change of understanding was identified. Understanding is a broad level behavior change but sufficiently specific for a program objective.

Learning Experiences. Learning experiences are those educational experiences which contribute toward the attainment of the educational objective. For example, a learning experience may be attending a meeting, going on a tour, or reading a circular letter. A learning experience might be seeing a demonstration on television or developing a fertility program using a case example during a meeting. In any instance, educational experiences or learning experiences are those things in which the *learner* engages so that he can change his behavior in a way that *both* he and the teacher see as desirable.

Learning experiences and teaching methods are closely related. The purpose of thinking about learning experiences is to keep the focus on the learner and his active involvement in the learning process rather than on the teacher.

Evaluation. Each element of a plan should include some references to possible methods of evaluation. Not every segment of every program must have a comprehensive evaluation annually. However, the program of instruction should indicate what will be evaluated, what evidence will be collected, and what techniques will be used to collect evidence. (See chapter ten for more specifics on evaluation of programs.)

Other Elements of Written Plan. Although the three most important elements of any written plan include objectives, learning activities and evaluation plans, there are other elements of a written plan which the teacher of adults will find extremely useful. These elements include: (1) methods to be used, (2) time of meeting or event, (3) place of meeting or event, (4) teaching resources needed, and (5) who will do the teaching. The written plan is not an attempt to develop lesson guides. It is merely an attempt to clearly identify the goals of any adult group, whether it be formally or informally organized. The following examples illustrate the types of topics included in programs for specific audiences. The objectives are not stated behaviorally in most cases.

EXAMPLES OF POSSIBLE PROGRAMS

For purposes of discussion, adult education programs can be classified under five different types: (1) Farm business planning and analysis programs; (2) Adult classes on specialized topics; (3) Agri-business management schools; (4) General adult farmer classes; (5) Informal educational endeavors.

Farm Business Planning and Analysis Programs. A formally-organized program of farm business planning and analysis has been developed over the past few years in Ohio. The vocational agriculture teachers conducting such programs built instructional units over a three-year period by working with a small number of farm families. They have received good response from the young farm families in this approach. When full-time teachers of adults are available, these families are requesting units of instruction above and beyond the first three years. Following is a listing of the instructional units for each of the three years in the Ohio program (7, pp. 6–8).

First Year—*Farm Records*:

1. The value of and the need for farm records
2. Uses for farm records
3. Selecting complete farm records
4. The importance and value of farm inventories
5. Taking the beginning inventories of non-depreciable assets
6. Taking the beginning inventories of depreciable assets
7. Using the cash account book-receipts
8. Using the cash account book-expense
9. Preparing the net worth statement
10. Developing farm business planning and analysis filing system
11. Introduction to federal income tax forms and calculating investment credit
12. Computing capital gains and/or losses
13. Keeping farm accounts current and accurate
14. Keeping household and personal accounts
15. Mid-year record book check
16. Review previous year farm business analysis report

Second Year—*Business Summary and Analysis*:

1. Making the ten-month summary
2. Income tax estimate and management
3. Enterprise accounts-livestock
4. Enterprise accounts-crops
5. Determining end-of-year inventories-closing out the farm inventories

6. Closing the cash account book
7. Completing the farm business summary form
8. Determining income and social security taxes
9. Analysis of net worth statement
10. General interpretation of first-year farm business analysis
11. Analysis of the size of business
12. Analysis of financial results
13. Analysis of the cropping program
14. Analysis of the livestock program
15. Analysis of labor, power, and machinery efficiency
16. Review accuracy of previous year's records

Third Year—*Adjustments in the Farm Business*:

1. Making the ten-month summary
2. Income tax estimate and management
3. Summarizing enterprise accounts
4. Closing out the five-year inventory book
5. Closing out the cash account book
6. Preparing the net worth statement
7. Completing farm business summary forms
8. Analysis of net worth statement
9. Completing income and social security tax forms
10. General interpretation of second-year farm business analysis (Identification of strengths and weaknesses)
11. Determining farm and family goals
12. Lowering operating costs (Operating ratio)
13. Reducing overhead costs (Overhead ratio)
14. Increasing labor and management income (Net margin)
15. Determining the most profitable cropping system through use of unit budgeting (Efficiency and volume)
16. Determining the most profitable livestock program through use of unit budgeting (Efficiency and volume)
17. Increasing gross income per $1,000 invested (Opportunity costs)
18. Adjusting power and machinery investments to increase net profit
19. Adjusting building and improvement investments
20. Adjusting labor to increase gross income per man equivalent
21. Developing and adjusting financial plan and debt repayment schedule

Adult Classes on Specialized Topics. Following is a brief discussion of a few specific programs conducted throughout the country.

The fact that there are only a few examples of production agriculture programs does not indicate that these were of less importance, but does indicate some other possible types of programs that are within this area.

J. W. Clark, Dane County, Wisconsin, agricultural agent, reported a successful City-County water resources program (4, pp. 11-12). This program was designed for the decision-making public officials in Dane County. Topics included in this program were: (1) What water resources do we have? (2) What is happening to our water resources? (3) How can we insure good quality in our lakes and streams? (4) A closer look at our water quality management problems. (5) What about legal governmental, financial organization of our water resources? and (6) How can we get the most from our water resources?

Clark brought in resource persons to help conduct the program, including a geologist, a professor of soils, a civil engineering professor, a professor of law, and a wildlife ecology professor.

Richard Travis, Area Extension Agent in Charge in Akron, Colorado, reported on an adult education program entitled "Community Business Institute," which was conducted for the small businessmen and their employees (14). Some of the topics discussed in this program were: (1) trends pertinent to retailing, (2) human relations, (3) consumer psychology, (4) selling, and (5) selling techniques.

County Agent John L. Sears in Safford, Arizona, reported on a program for cotton growers (13, p. 11). This program for cotton growers included the following topics: (1) water measuring devices, (2) preemergence herbicides for control of weeds and cotton, (3) weed control in cotton seed, (4) rust damage in cotton, (5) use of long staple strains, and (6) field crop alternatives for cotton.

Muskrat control was one element of the program for rice producers in Loenoke County, Arkansas, according to Bobby A. Huey, County Extension Agent (6, p 1). Agent Huey reported that the muskrat meeting, as one of a series of meetings and other direct field teaching, was one of the best meetings held in their county in recent years, with over 90 farmers attending.

These examples demonstrate the variety of topics that can be geared toward specialized audiences.

Agri-business Management Schools. Turf management landscape owners were the target audience of vocational teacher Vincent Feck, in the Cleveland, Ohio, Public Schools (5). In this horticultural adult class the following topics were discussed: (1) economic principles affecting management, (2) inventorying resources, (3) management goals, (4) income, (5) maintenance programs, (6) labor, and (7) fiscal records.

General Adult Farmer Classes. Frequently these classes grow from the formally-organized group as a part of a vocational agriculture young farmer or adult farmer class. In many instances, young farmers are brought together not because of their specialized common agricultural interest, but because of their common interest in developing leadership within the community and because of the social benefit of meeting with other young farmers in the community.

It is difficult to find a specialized topic that fulfills the educational needs of all members of such groups, unless it is in the area of farm business planning and analysis. (For example, a program on beef cattle production would not be possible if there were a number of dairymen or swine producers in the group.) Thus, many times the general adult program is one which serves common needs. Typically this type of program will include such topics as: farm outlook, farm record-keeping, income tax analysis, welding classes, and topics related to the production and marketing of specific products or services. Such programs help to develop leadership and social abilities, thus being highly useful even though some of the members of the class may not always have a topic directly appropriate to them.

Informal Programs Outside of the Organized Meeting. Mass media is one of the major tools used by Vivian N. Jennings, Extension Leader, Crop Production in the Cedar Rapids office of Iowa State University (9), as he conducts his crop production program. He recently completed radio and television series including two different six-program series on corn and soybean production, a three-program series on chemical safety, and an eight-program series on understanding soil fertility. Jennings worked with both radio and television to reach his agronomy audiences. In addition, he conducted meetings and clinics with the audience in his area. Jennings also writes crop production information letters dealing with current topics and sends them to top farmers, agri-business leaders, county extension directors, vocational agriculture instructors, and the mass media in the area. Parts of one of his newsletters are reproduced below to illustrate his approach (8):

> *2-4-D . . . . 'Buggy Wipping'*
> Reports of 'Buggy Wipping' in corn where 2-4-D was applied have been coming from several localities. The problem usually occurs when 2-4-D is applied to corn six to eight inches high. However, it seems under certain conditions and susceptible varieties, it can occur when corn is smaller than this. Application of 2-4-D beyond third or fourth leaf stages is risky business.
> *Soybean root rot.* Root rots on soybeans continued to cause problems. In many. . . .

Educators should not neglect these informal approaches to teaching. Many audiences may prefer this method of learning to that of being in a formally-organized course. In a Pennsylvania study (3), the newsletter was preferred as the method to receive agricultural information over six other mass media and individual teaching methods.

Process of Planning a Program

Adult agricultural education programs will exist only if they are designed to meet the *needs* and *interests* of our clientele as they see them. Thus, a sound process of program development must be followed and the potential learners must be involved in making decisions about the program's objectives.

Keeping in mind that in adult education the potential students should be involved in the planning process, we shall examine some basic planning procedures.

Tyler cites four fundamental questions concerning the design of any curriculum (15, p 1). These are: (1) What educational purposes should the adult educator seek to attain? (2) What educational experiences can be designed to achieve these purposes? (3) How can these educational experiences be effectively organized? (4) How can we determine if the intended directions have been achieved?

Another way to look at the basic curriculum design questions is shown in Figure 5—1.

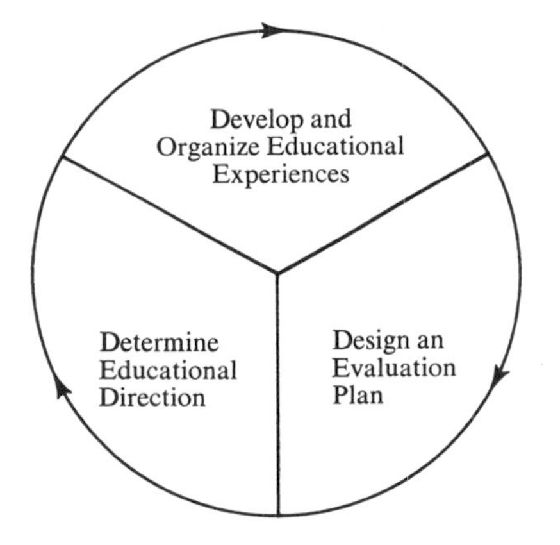

FIGURE 5—1 PROGRAM DEVELOPMENT PROCESS

This model indicates three distinct phases in curriculum building:

1. Determining educational direction or objectives
2. Developing and organizing educational experiences
3. Designing evaluation plans

Tyler has carefully broken each of these elements into more distinct steps in program planning. After presenting one other figure explaining a program development model, we shall explore the first step—determining objectives—in more detail.

Pesson has identified a planning model which is another useful way of looking at program development (12, p. 95).

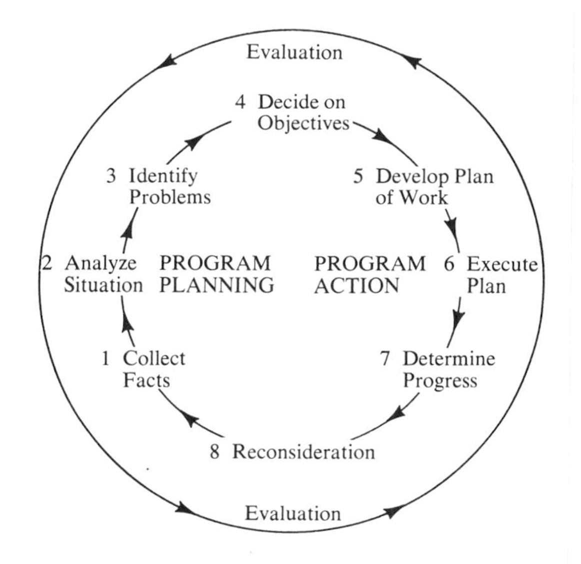

FIGURE 5—2 PROGRAM DEVELOPMENT PROCESS

The model as presented by Pesson involves eight fairly distinct steps in *planning* and *action*.

1. Collection of facts
2. Analysis of the situation
3. Identification of problems
4. Decisions on objectives
5. Development of the plan of work
6. Execution of the plan of work
7. Determination of progress
8. Reconsideration

This conceptualization of the planning process is different because it also involves the action phases of the program. Tyler's basic questions refer only to the planning (including the planning for evaluation) and not action. For successful program development, a mental process must be followed which can help a group of lay people to reach sound decisions on program direction.

DETERMINING EDUCATIONAL DIRECTIONS

The following ideas are useful in establishing educational direction:

1. Collect and analyze data regarding a problem area that will help to establish a clear understanding of the present situation.
   a. Local situational data
   b. National and international influence
   c. Trends in area of concern
   d. People's interests and needs
2. Collect and analyze data regarding a problem area that will help to establish the desirable objectives of the program.
   a. Research findings
   b. People's interests and needs
3. Establish the most important needs of our given problem areas through discussion and interaction with the potential learners. This might be conceptualized as in Figure 5–3.

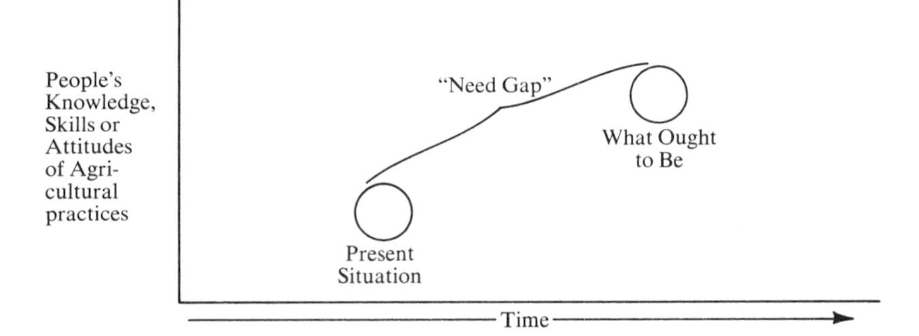

FIGURE 5—3 THE "NEED GAP"

The idea to be conveyed in Figure 5–3 is that we must clearly understand the present situation in regard to people's knowledge, attitudes, or skills or their adoption of practices. The extent to which this present situation differs from what ought to be—as determined or viewed by the learner—will indicate the needed educational effort.

As an example, in a dairy community, records indicate the present level of milk production at 9,000 lbs. Knowing that 12,000-14,000 lbs. levels are easily attainable (several dairymen are now at this level) the planner must look at practices being used and the farmers' level of knowledge, attitudes, or skills which may be affecting this level of production. For instance, he may find that poor breeding practices are causing low production levels. Other possible causes of low production may be poor feeding, inadequate housing, or insufficient credit for expansion. Because one program cannot solve all concerns in one year, the planner must decide on the most important problems, concerns, opportunities, or objectives. Kirby has graphically presented one approach to this decision-making task (10, p. 15).

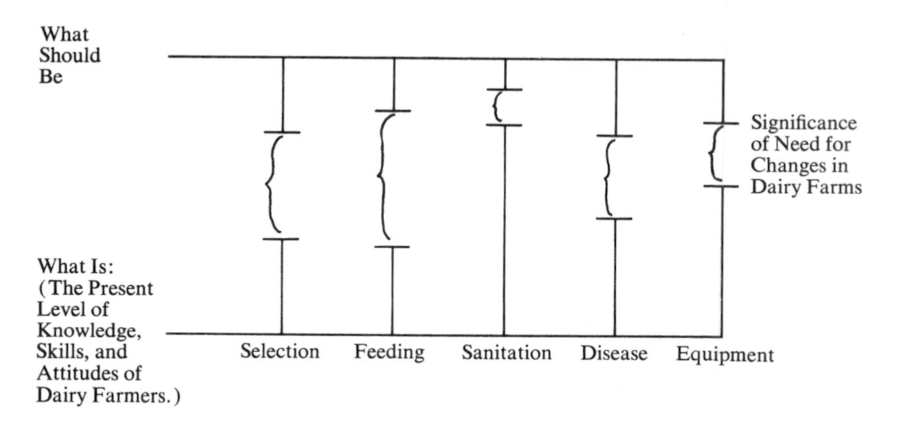

FIGURE 5—4 ESTABLISHING PRIORITIES

Figure 5–4 illustrates that by clearly identifying the present situation in each area and recognizing that each area is not equally important in a total enterprise, the planner can establish priorities. Thus, in Figure 5–4 it is shown that feeding may be the most important problem in this dairy community. Further analysis must be made of the farmers' actions and knowledge as related to the feeding inadequacies. For example, the farmers may not know about a new product which reduces costs and increases profits by including more urea in the ration. Or perhaps the farmers have not mastered a system for extra feeding of the high producers, which cannot eat an ample amount in the milking parlor.

In summary, to analyze the present situation and the ideal situation, the planner must examine practices used or not used and explain how this use or disuse affects maximum efficiency or effectiveness.

The ultimate outcome of program planning is the expected objectives of the program. An educator can teach many subjects without having a clear conception of his objectives. However, a clear statement of his objectives—preferably a written statement—will usually make the educator more effective. There are three main levels of objectives:

1. Broad organizational objectives
2. Program objectives in specific areas of work
3. Teaching or instructional objectives for a specific lesson to be taught

The first level of objectives might well include such statements as:

1. Farmers to improve their standard of living in the community as a result of improved agricultural practices.
2. Agri-business leaders to better serve the needs of commercial agriculture through community development programs.
3. Farmers to increase production of alfalfa per acre to 10 tons.

These types of statements are rather broad but serve to identify the audience and the general subject matter area.

It is at the second, or program objective, level that the preceding discussion has been directed. In specific fields or enterprises, objectives aid the teacher to clearly identify: (1) the specific audience; (2) the desired behavior to be developed as a result of the program; and (3) the specific content areas that will be taught. These are frequently referred to as the three elements of objectives—audience, behavior change, and subject matter.

The following are two examples of possible program objectives for the dairy problem discussed earlier:

1. *Farmers* to *understand* the *benefits of using new products* in dairy rations.
2. *Farmers* to develop the *skill* to design a *barn arrangement and feeding plan* to permit higher producers to receive extra feed.

In each of these examples, the three elements of objectives—audience, behavior change, and content—are in italics in that order. Any good objective should clearly identify these three elements. Some other criteria for good objectives include:

1. Will they lead to useful outcomes?
2. Are they likely to lead to action?
3. Are they important to large numbers of people?
4. Will they lead to economic development?
5. Are they socially desirable?
6. Are they attainable through education?
7. Can they be evaluated?

The third level of objectives is to develop teaching or instructional objectives for specific lessons. At this point, the teacher will need to be more specific about the audience, the content, and the desired behavior change. Mager (11) is a good source for further discussion of instructional or teaching objectives.

The program planning process is a deliberate thought process. Educators must engage in interaction with people, which includes a careful analysis of the situation as reflected in: (1) people's needs and interests; (2) societal needs; (3) research; (4) local facts.

Although the situational statement is normally not written out in detail, the following is an example of a situational statement for a county with a dairy problem. This statement is followed by the statement of some objectives that might well be derived from this analysis.

County Situation. General County is one of the more highly productive agricultural counties in the midwest. Gross income from agricultural production normally runs just below 20 million dollars annually. The county also has a fair amount of other industry.

Dairying is the largest source of agricultural income, accounting for 60 per cent of the total. There are 1200 producers ranging in size from those having only a few head to a number of farms having 100 to 200 producing cows.

There are 4800 cows on official dairy herd testing programs. The average production of those on test is 12,056 pounds of milk and 456 pounds of fat. The average production estimate of all dairy cows in the county is around 9500 pounds of milk and 375 pounds of fat. The county average is above the state average of all cows, but of cows on test in the state, it is just average.

The county agent and vocational agriculture teachers feel that two types of problems are prevalent. There are a number of marginal producers who have poor breeding and nutritional practices. Most high-producing cows are not being fed ample nutrients while in the milking parlors.

A number of progressive farmers have expressed concerns about prices for milk, marketing procedures, and labor problems. The marginal producers have not expressed much interest in any type of program.

Analysis of Situation. A number of problems may be evident from the above example.

a. Production is low for a high-income dairy county.
b. Marginal farmers show little interest in education even though they have many nutrition and breeding problems.
c. Nutritional adequacy is a problem with high-producing herds.
d. Marketing is a concern, especially as it pertains to adequate prices.
e. There is interest in ways to cut labor needs and costs.

Educational Objectives Which May Be Appropriate. Three examples of educational objectives are given below:

1. Producers to develop an understanding of dairy marketing procedures and problems.
2. Producers to develop ability to analyze housing and feeding practices to cut labor.

3. Small producers to develop an understanding of improved breeding programs.

These are but a few objectives that might be developed from the above situation. The important thing is that we have a systematic procedure to come up with program objectives.

Student Involvement in Planning

If a program is to be successful, the students must be involved in its planning. Learner involvement in the planning process has several distinct advantages:

1. Students will perceive a program as more valuable if they have been involved in the planning. (In most cases, both the educator and the learner will see it as a better program.)
2. The learners who are involved in planning will aid in supporting programs through attendance and recruitment.
3. The program will become "our" program, not the teacher's program.
4. Their involvement helps the learners to understand and support the entire program.

Some disadvantages do result from involving potential learners in planning. Some of the more critical are:

1. It takes more time to plan.
2. It is more difficult to involve others.
3. The program may not be exactly what the educator had planned.
4. Planning groups tend to apply more pressure to work more on *their* programs.

In spite of these problems, it is essential that the educator use either an individual or group approach to planning with the learners in adult education.

For years, the procedure of working with key individuals has been used and mastered in adult education in agriculture. In this approach, the educator makes individual contact with three to five key farmers (or other audience groups) to see what they are doing and what ideas they may have for a program. Some use of this technique will never be replaced, even with extensive committee planning efforts.

More recently, planning committees have been used to plan adult education programs. This practice has probably developed as the result of (1) a better understanding of the nature and benefits of small group action, (2) an effort to make more effective use of time, and (3) the need for programs with many different audiences rather than just a single adult farmer audience.

Two types of adult education planning committees are generally used in agriculture; namely, special interest and overall advisory committees. The most frequently used is the committee which plans a specific program, such as (1) young or adult farmer multi-topic; (2) swine; (3) dairy; (4) agri-business; or (5) horticulture; for a specific audience. This committee generally has little concern for other adult education programs in progress.

The other type of adult education committee is the overall advisory committee, which has been used more frequently within the Cooperative Extension Service because of the many specific commodity committees in existence. This type committee functions to look at *all* the programs of adult education in a community or county. With the expansion of adult education programs in vocational agriculture, it may become increasingly important to have one group that focuses on all the adult education programs in a school district to see that the most important problem areas are being covered.

Although the overall advisory group is important, the major discussion which follows is directed toward specific interest planning committees. Before looking at the structure or organization of a committee, the educator and the committee members must establish the committee's function.

FUNCTIONS OF PLANNING COMMITTEE

A planning committee might perform one or all of the following general functions of committees: (1) Advisory, (2) Decision Making, (3) Informational, (4) Public Relations, (5) Legitimizing, (6) Action. Each of these functions will be reviewed as it relates to adult education planning committees.

Advisory. Sometimes a committee advises the educator regarding program needs as its only function. However, most people who consent to serve on planning committees hope to contribute more than advice; they hope to see results from their deliberations. Most educators do give advisory committees additional functions. However, the important point is that if committee members are to be advisors, they must be made to understand that that is their *only* function.

Decision Making. Most specific interest planning committees do help make decisions about program content, for they are involved in the resulting program, but they normally leave the development of specific educational experiences to the professional educator.

Boyle in his summary of research findings on program development has indicated that "data that are available and used by planning committees tend not to be adequately analyzed or interpreted" (2). This implies that educators want committees to make decisions, but seldom give them time to make wise decisions. If committees are to make decisions, they must understand this function and have time to digest the information regarding a problem or situational area.

Informational. The informational function is reciprocal. The committee may bring information about the community to the program planners or it may receive information about a program for its own consumption and/or use in another role of public relations.

Public Relations. A public relations committee has the task of seeing that the community knows about the adult education programs in agriculture. It may focus its efforts on potential learners or on helping to inform key people such as a school board, county commissioners, school executives, business leaders, and government officials about the program.

Legitimizing. Although the legitimizing function may be seen by some as a "rubber stamp" effort, it is really greater than that. In their book (1, p 81), Beal and others have defined legitimation as giving sanction (authority, approval, or justification) for action. They point out that there are both formal and informal legitimation structures. A planning committee might well serve as the formal structure—but the good educator knows that for large community projects the informal legitimation structure (composed of leaders with power or influence) must be considered. The educator must remember that many legitimizers may not be involved in the actual program; thus, he will need different types of committees on major projects.

Action. Planning committees create action by doing considerable work beyond initial planning or public relations. Committee members often help with the physical and mental arranging for educational programs; they teach, either individually or in small groups, they arrange tours, plan special activities, and see that needed resources are available.

For most programs of adult education, a planning committee performs all of the six mentioned functions: advisory, decision making, informational, public relations, legitimizing, and action.

STRUCTURE OF PLANNING GROUPS

The structure of a planning group should relate to the functions it is to perform. Since a committee usually performs all the functions mentioned above, some guidelines will be helpful in structuring a planning committee.

Membership on a Committee Should Be Representative. A program's audience and subject areas must be specified before a committee is organized. If the committee is to plan a total adult education program, all agricultural enterprises should be represented. If the program is specific, e.g., one for elevator employees, these employees and/or their managers should comprise the committee. Not only all groups, but also all geographic areas affected, should be represented if the program is to be successful.

Members Should Represent Those Who Selected Them. Committee members cannot truly express the needs and interests of the people if they represent only themselves. They should feel a responsibility to bring in ideas of others and to take back programs which were agreed upon. When the committee member thus involves his constituents in the planning of the program, they are more likely to feel a part of the program and to support it.

Members Should Serve Long Enough To Make a Contribution. The length of time required for making a contribution varies, of course, depending upon the individual and the problem at hand. However, for true group interaction to occur, committee members should serve for at least three to five planning sessions.

Some Rotation Plan Should Be Followed To Keep Committees "Alive." The periodic addition of new members keeps a group from becoming stagnant. It is well to establish at the beginning whether a committee will be (1) a continuing one, or (2) an ad hoc one. If the former, the rotation plan needs to be clearly identified at the start to avoid hard feelings when a change is needed.

There are other guidelines to effective committee structure, but if the preceding are followed as *interested* members are selected, effective planning committees can be established.

REFERENCES

1. Beal, George, *et al., Social Action and Interaction in Program Planning.* Ames: Iowa State University Press, 1966.
2. Boyle, Patrick G., *The Program Planning Process.* Madison, Wis.: National Agricultural Extension Center for Advanced Study, 1965.
3. Brown, Emery J., "The Evaluator," University Park, Pa.: Cooperative Extension Service, 1968.

4. Clark, J. W., *Let's Look at Our Water Resources.* Madison, Wis.: City of Madison—Dane County Water Resources School, 1966.
5. Feck, Vincent, "Turf Management Landscapers Course Outline," Cleveland: Cleveland Public Schools, 1968.
6. Huey, Bobby A., "Loenoke County Monthly Report." Loenoke, Ark.: Cooperative Extension Service, 1968.
7. "Instructional Units for Teaching Farm Business Planning and Analysis in Ohio." Columbus: Agricultural Education Service, 1969.
8. Jennings, Vivian N., "Crop Production Information Letter No. 5." Cedar Rapids, Ia.: Cooperative Extension Service, 1968.
9. ———. Letter to Author. Cedar Rapids, Iowa, 1968.
10. Kirby, E. L., "Developing More Effective Extension Educational Programs," in *Program Development Workshop for County Extension Agents.* Columbus: Cooperative Extension Service, 1963.
11. Mager, Robert F., *Preparing Instructional Objectives.* Palo Alto, Calif.: Fearon Publishers, 1962.
12. Sanders, H. C., *The Cooperative Extension Service.* Englewood Cliffs, N. J.: Prentice-Hall, Inc., 1966.
13. Sears, John L., *Serving You—Graham County Annual Report.* Safford, Ariz.: Cooperative Extension Service, 1966.
14. Travis, Richard L., Letter to Author. Akron, Colo., 1968.
15. Tyler, Ralph W., *Basic Principles of Curriculum and Instruction.* Chicago: The University of Chicago Press, 1950.

Chapter 6 TEACHING ADULTS IN CLASS

Introduction

The professional educator recognizes that teaching is both a science and an art. One of his tasks is to learn the separate characteristics of the art and the science of teaching and then skillfully to combine these elements to create an effective teaching-learning situation. The utilization of the science and the art of teaching by the professional educator is quite similar to the preparation of a formulation for a prescription by an expert pharmacologist. However, too often the science and the art of teaching are oversimplified or misunderstood by the public, and professional educators should promote enthusiastically the proper preparation of all teachers. Otherwise, teachers may enter the field with the impression that only fortitude and knowledge of the subject are required for teaching adults. In this chapter, attention is given to both the science and the art of teaching with directions for their employment so that the teaching-learning process may be enhanced.

The Science of Teaching

Chapter 3 presented an organized body of educational knowledge applicable to teaching. Some psychological and sociological concepts and known teaching-learning relationships were identified. The importance of the science of education should not be minimized, but at the same time, its function in education should be recognized. This is clearly indicated by William James, Talks to Teachers.

> I say moreover that you make a great, a very great mistake, if you think that psychology, being the science of the mind's laws, is something from which you can deduce definite programmes and schemes and methods of instruction for immediate classroom use. . . . sciences never generate arts directly out of themselves. An intermediary inventive mind must make the application by using its originality.

The Art of Teaching

The creating and/or manipulating of situations to secure the desired student response involves method and skill. As Lyman Bryson (1, p. 11) points out:

> . . . whenever one accepts responsibility for teaching, he has need of method. If all those who know what others would like to know were geniuses in communicating thought, there would be no need for systems or pedigogy. Method is the only substitute for divine inspiration.

Ability to use method distinguishes the competent, genuine professional from the amateur. By controlling the teaching situation, the genuinely professional teacher makes the learning results predictable. This chapter emphasizes some methods, with skills and understandings, that the teacher should use to bring about learning by adult students.

EDUCATIONAL NEEDS OF ADULTS

The educator preparing to teach a group of adults should consider adults' educational needs such as social, civic, family living, health, ethical, avocational and vocational. This presentation is primarily concerned with teaching adults as it relates to their vocational needs in agriculture. But it should not be inferred that teachers are to eliminate or evade the other, non-vocational, educational needs of adults. Rather, they should relate the non-vocational and vocational needs when the occasion arises, to strengthen the bonds of learning.

CONSIDERING INDIVIDUAL DIFFERENCES

In adult classes, the common agricultural needs of the individual members should constitute the program of instruction. However, since all the problems of every student cannot be solved by class instruction, there is also a place and a reason for individual instruction. Such teaching could conceivably be provided before or after regular class sessions, at the place of business, farm or home. The emphasis in Chapter 6 is on teaching adults in classroom situations, while in Chapter 7, supplementary methods such as individual instruction and small group instruction are presented.

Teaching the Class of Adults

Teaching adults in classes is in some ways similar to teaching high school students, yet there are differences. Obviously, teachers of youth who are preparing to teach adults will have adjustments to make in procedure. In the succeeding pages some suggestions are given to assist the teacher of adults in perfecting his teaching techniques.

Lyman Bryson provides an orientation for aspiring teachers of adults in the following excerpt (2, p. 79):

> All methods in adult education are more or less formal. Every group is made up of people differing in experience and in preparation. The average mature person will dislike the regimentation of the old fashioned class, even though he may be docile to the authoritarian teaching. Informality is necessary.
>
> It is not possible, however, to relate classwork on the informal methods exactly to the functional types of adult education. In general, we may say that regular classes will be used for the most part in remedial work. Classes with the shop as a laboratory will be the basic method in occupational training. The discussion group is perhaps the most useful form for organizing relational work. In political education, the lecture, the discussion group, and the forum, which is a combination of the two, are generally relied on. In all work with adults, all methods should be put into practice if they offer any advantage for a special phase of the student's experience.

Bryson says, in effect, that most adult education is informal because students differ in preparation and experience. Consequently, formal class sessions are not only disliked, but are usually unproductive. Bryson does recognize formal procedures are necessary in teaching functional classes such as vocational agriculture. In this situation, during the regular class sessions attention would be given primarily to remedial work, such as helping students to clarify difficulties, locate information, evaluate facts and apply decisions. He does not mention the techniques for employing discussion groups. For teachers of agriculture, a very effective and most frequently used procedure is problem solving. Other group discussion techniques, such as forums, panels, symposiums and buzz sessions are also available to teachers of adults.

Bryson states that for occupational training, in addition to group discussion, a laboratory is needed. This could include a school laboratory, a land laboratory, a school farm, a student's farm, an agricultural business, public lands and in fact any place or combination of places where participating experiences may be secured for students.

Bryson discusses the lecture method as used in adult education in the following (2, pp. 80-83):

> If lecturing is bad, nothing can be said for it. If it is good, it is mostly a stimulus. When the stimulus leads to further intellectual effort, the effect is something worth having. The wise educator tries to build from these things the after-efforts which make them genuinely educational.

In general, Bryson says, all methods which offer an advantage in the teaching-learning situation should be used in adult education. He does not rule out the lecture, but includes it with other methods.

An action-centered vocational education program may be more effective in helping adults to improve their vocations, since it often results in improved practice and a better living. Paul Sheats and others in adult education say that (7, p. 323):

> . . . adult education programs should be oriented to meet the felt needs of people and that these felt needs are generally satisfied by methods designed to pass subject matter information efficiently from authorities (the text books, the reading references, or the instructor) to the student. Felt needs are more likely to be satisfied by methods which involve the use of scientific method—that is the identification of the problem or the block to needs satisfaction, the gathering of data, the formulation of a hypothesis, and finally the testing of this hypothesis through an action program. Evidence is accumulating which indicates that problem-solving, including the actual bringing about of individual or social change, can be best done most effectively not by isolated, individual effort, but rather by the cooperative participation of groups of people.
>
> If these points of view are accepted—that is, first, that adult education should meet the needs of people, second, that meeting needs generally demands problem-solving, third, that problem-solving involves

> action programs, fourth, that problem-solving, including the promotion of action, is best done through group participation, then it follows that the fundamental problem in adult education method is not concerned with formal teacher-dominated activities such as effective ways to lecture or demonstrate or to make assignments but rather with the problem of setting the stage and providing leadership so that groups may most effectively work toward the solution of their problems and the satisfaction of their needs.
>
> Cooperative group discussion as a method of learning casts the teacher or leader in the role of a co-equal member of the group, bearing a responsibility with others in the group of bringing personal knowledge and experience to bear on the problem or areas which the group has selected for study. As an educational procedure, the method is based upon the principle of learning which recognizes that a person holds more enduringly and more intelligently those concepts which he has participated in formulating (7, p. 328).

Problem solving is not particularly new to teachers conducting classes with adults. However, it has often been misused. One of the patriarchs in agricultural education is Carsie Hammonds of Kentucky, who has this to say about using the problem solving approach (6, pp. 78-79):

> Most good teachers of most subjects do their teaching partly or wholly on a problem basis. They guide or direct their students in solving problems rather than attempt to teach facts from textbooks.
>
> Problem solving is a thinking process. Thinking by the students is usually necessary in good teaching, the essentials of method being the essentials of reflection. Problematic situations for the students must exist if the students are to think reflectively; they should be used in teaching if we expect the students to do reflective thinking while being taught.
>
> The use of problems in teaching should play an important part in good vocational preparation. Many vocations consist, to a great extent, of solving problems. To the extent that it deals with unstable and changing situations, every vocation has its elements of newness that cannot be effectively handled by fixed habit. Hence, there is a difficulty, alternatives to be evaluated—essential elements of a problem. Experience in problem solving is the primary if not the sole means of developing adaptiveness. Sensitivity to problems is a major characteristic of creative work.

From the foregoing, it is quite obvious that when the discussion method is employed effectively, the teacher can help students in solving their problems. In the succeeding pages some suggestions are given to assist teachers in organizing instruction for adults so that their vocational needs are fulfilled.

Preparing for Class Instruction

IDENTIFYING THE TOPIC

The subject for consideration by the class is taken from a program of instruction, where it may be stated as a topic, a group problem, a specific problem or may be included as a part of a broad subject. For example, the curriculum for a course could identify a subject as *Selecting the Variety of Soybeans Needed for Grain Production,* or *What Variety of Soybeans Should Be Selected for Producing Grain in Our Community,* or *What Variety of Soybeans Should John Miller Select for Grain Production* or *Growing Soybeans Profitably.* Whichever procedure is followed in the program of instruction, the teacher should properly title the teaching content so the adult students know what is to be taught. A properly worded title has appeal and may be an aid in securing attendance or in contributing to the public relations program.

STATING A PROBLEM

Some teachers follow a system in stating a problem. They use the adverb *how* to denote the entire problem, and not to denote *what method* or to state a segment of the major problem. In such a system, such words or combinations of words as why, what, which, where, what amount and what method are used to begin the statements of subordinate problems.

The verb *should* is most frequently used in the problem statement, in preference to such words as shall, will, could, might, may, ought and can. *Should* denotes what is desirable, but does not command what is to be done.

DETERMINING SPECIFIC OBJECTIVES

One of a teacher's most difficult tasks is to formulate meaningful teaching objectives. He can easily be confused, since the adults' many needs can be associated with the lesson topic in various ways. He knows that there are not only such needs as social, religious, health, family living, avocational and vocational, but in each there are involved ideals, attitudes, appreciations, interests, understandings, and abilities. Consequently, the teacher can go easily from one extreme to the other; he can prepare a compendium of objectives that are almost senseless, or he can attempt to teach without the benefit of clearly defined guides.

Objectives should be confined to specific, desired outcomes from the teaching. Using the example, "What Soybean Variety Should John Select?" the one specific objective is indicated in the problem: *Select a Variety of Soybeans.* Since the teacher knows the problem, and how to solve the problem, he should also have arrived at a correct decision about which variety will meet the objectives of the lesson. Certainly all problems do not have only one objective, but the teacher can identify meaningful, purposeful, achievable, and measurable objectives by carefully stating problems or lesson topics.

Using the example, "How Should John Meet the Lime Requirements for Growing Alfalfa?" the teacher is concerned with student involvement in such objectives as these:

1. Determine the pH of the soil in the field set aside for growing alfalfa. (If this service is not available at a reasonable cost, the students should be taught how to take soil samples, the method of testing for pH, and where to send the samples for testing.)
2. Determine the pH requirement for growing alfalfa on the soil in the field set aside for the crop.
3. Determine the amount of the lime needed for the kinds of lime available.
4. Determine the kind of lime to be applied.
5. Determine when the lime should be applied.
6. Determine the method of applying the lime.

THE IMPORTANCE OF OUTCOMES

Note that the above objectives need not be changed when using any one of the following topics: Meeting the Lime Requirement for Growing Alfalfa, Liming Needs of Alfalfa, and Lime and Alfalfa. The difference is in method. Regardless of the method used, if the teaching is effective, the students' changes in using lime for alfalfa should be consistent with the teacher's objectives. The teacher would say that his success is measured by the effected favorable change in the students' behavior, while the layman would say the teacher has taught when the students follow the instruction.

WRITTEN OBJECTIVES AND OTHER TEACHING PURPOSES

The instructor is expected to include the students' other appropriate needs in the lesson, and not to confine his teaching to the stated objectives. Sometimes he may incorporate notes in his lesson plan to remind himself when to deviate. However, the teacher must be careful not to neglect the important items when he is presenting the less important ones. Lesson objectives have a real function in the teaching procedure, and, when properly used, they give direction and keep the teacher and the class on target.

Preparing The Teaching Outlines

PREPLANNING AND SIMULATING

Before he begins to teach, the teacher should preplan the procedure he is going to use, simulating the experiences he anticipates during the class session. In so doing, he should be alerted to what is required to make the actual teaching experience a success. If he preplans well, he knows the subject, why it is important, the desired outcomes, and the difficulties that students will have in understanding the subject, in developing the needed abilities, and in adopting the practices. He will know his own inadequacies in understanding the subject and in teaching the doing parts of the objectives. He will know what on-farm instruction and consultation is required to get the individual students to improve their practices. He will know what aids to use and when to use them. In fact, planning and previewing should help make preparation for teaching less time-consuming and more meaningful.

REWRITING NOTES

In preplanning a lesson, some teachers find it convenient to keep a written record of the preview experience rather than to rely entirely on memory. This written record can be organized into a lesson plan or teaching outline. Since a lesson plan is so highly personal, its content is for each individual to decide. Its adequacy can be judged by the success of the teaching. If the teaching was effective, the preplanning and teaching outline probably sufficed. However, if the teaching was ineffective, then the preplanning and the teaching outline could have been among the many factors involved in the failure.

COMMON ELEMENTS IN LESSON PLANS

The many variables associated with the art of teaching, the differences among teachers, and the peculiarities of the many subjects to be taught make it difficult to formalize lesson planning. However, lesson plans do have common elements, and a study of procedures used should prove beneficial to beginners and others who are not satisfied with the system that they are using. Assistance may be secured by reviewing *The Student Teacher in the Secondary School,* by Lester and Alice Crow (4), and *The Experience of Student Teaching,* by John Devor (5).

The following partial list of teaching elements may be helpful to a teacher in formulating a plan:

Subject (or problem)
Teaching objectives

Subject breakdown in logical steps
References
Significant facts
Key questions
Incidents, illustrations
Aids
Special procedures
Assignments
Important things to learn [definitions, principles involved, procedures to follow in learning (including safety measures)]
Decisions and conclusions
Plans to carry out decisions and to secure experiences
Tests and other appraising devices
Evaluation of learning
Evaluation of teaching

These elements can be separated into two parts: one that is primarily for the teacher and the other for the students. In an arbitrary division, the units are related entities and are coordinates. It is helpful in making the division to include items only once, but to place under *student* those items that he needs to develop a complete set of notes. With this in mind, we could divide the list as follows:

Teacher	*Student*
	Subject or problem
Teaching Objectives	
	Subject breakdown in logical steps
References	
	Significant facts
Key questions	
Incidents, illustrations	
Aids	
Special Procedures	
Assignments	Important things to learn: [definitions, principles involved, procedures to follow in learning (including safety measures)]
	Decisions and Conclusions
	Plans to carry out decisions and to secure experiences
Tests and other appraising devices	
	Evaluation of learning
Evaluation of teaching	

THE OUTLINE FORM

The items in the outline may be recorded on a series of pages like a book, following a time and use sequence. Thus, both the teacher's part and the student's part are included in the plans. Other teachers prefer to use an open notebook and to separate the teacher's part from the student's part. In essence, they are taking page by page the outline first described and placing the teacher's items on the left page, leaving blank spaces on the right page for recording the student's items. The advantage of this system is that the materials to be placed on the chalkboard and to go in students' notes are isolated and clearly in view. The disadvantage is that for some teachers, the separation of the outline in parts has no particular value. Regardless of the system used, a good teacher plans for teaching.

FORMULATING THE PLAN

A busy teacher of adults has little spare time. He schedules his work so as to give priority to the things that are most important. Planning for teaching falls into this category. One of the better teachers, whose adults meet weekly, begins to develop his outline when introducing a new subject, the day after the previous session with his class. Of course, he has given thought to the subject and its presentation from the time it was placed in the program of instruction. So in this hour set aside, he reviews his thinking, as well as other outlines pertaining to the problem, and decides how he will teach the class. He states the objectives, segments the problem by breaking it down into logical steps, checks the references available, determines what additional materials are needed, sets aside one or two references for study when he has a few moments, and then prepares a list of things that he is to do with a schedule for getting them accomplished. With the problem of *Selecting a Variety of Soybeans* as an example, the list of things for a teacher of vocational agriculture to do could include:

1. Call at the farms of Harry Jones and John Fisher on Wednesday evening, and Henry Davis on Thursday evening, to determine exactly what their experiences have been with soybean varieties, and what they have learned from others about some of the newer varieties.
2. Stop at the elevator on Wednesday morning on the way to school and get the price lists and availability of the various soybean varieties, and also get literature, seed tags, and the elevator manager's opinion.
3. Call Wednesday during the fifth period (set aside for confer-

ences) the Area Extension Agronomist, the Certified Seed Grower, and the agronomist in charge of research in soybean varieties at the Research and Development Center to discuss the problem and its evaluation.

4. Give the school secretary a copy of the report, *Demonstrate Plot Tests with Soybean Varieties,* on Wednesday for duplicating.
5. Prepare charts of soybean variety tests Friday after school.
6. On Friday after school, set aside the references available for teaching.
7. Review during the week the recent publications for information on soybean varieties.

SCHEDULING TIME FOR PLANNING

Assuming that the previous class session was on Monday, teachers should begin planning the next session on the following Tuesday by starting the plan and determining what they need to do before Friday. Should there be several unfinished tasks, they could be accomplished on or before Monday, the day of the next session.

Some teachers find it very helpful to evaluate their class sessions in writing soon after the meeting and file these suggestions to assist in planning similar classes at a later date.

USING THE PLAN

After a teacher prepares a lesson plan or teaching outline, it should be an asset and not a liability in his teaching. The well-prepared preview should provide the clues for staying on the subject and meeting the objectives. The teacher should go over his plan as often as necessary so that at the time of the meeting he is thoroughly familiar with its content. This plan in its proper order should be placed on the lecturn in the classroom for ready reference. Picking it up from time to time is not particularly distracting to the class if the teacher can get what he needs by a momentary glance at the outline. The teacher may read from the outline significant exerpts from authorities on the subject, if they add to the discussion or presentation. In fact, this system may be easier and more effective than reading the quotes from a collection of specific references and using them in the procedure.

Often the misuse of an outline gives a negative reaction to lesson planning. Such reaction may be an excuse by those who do not plan well or have not learned how to use plans. Regardless, it should be remembered that the good teachers do plan and they use their outlines when teaching.

Teaching The Class

GETTING STARTED

The teacher should have planned carefully his approach to the class. If he has been teaching the group, he will have established the introduction. If he is new to the group, getting acquainted with some or all of the students before the first class session will help. Working with an advisory committee assists to break down some of those real and/or presumed barriers that affect teacher-student relations. Whatever the situation, the teacher's task is to get the learning process started. He is to create and/or manipulate the learning situation so that the student wants to do what he should do. When the teacher does his part properly, he gets favorable student reaction.

The meeting should be started according to the arranged schedule. It is the teacher's responsibility to keep this schedule. The tone of the meetings, the students' attitude toward tardiness and, to some degree, absence are affected by what the teacher says or does. If he starts on time, has a real problem, moves the discussion logically, arouses interest, reaches worthwhile objectives, and stops on time, he will succeed. He will have the qualities associated with good teachers. Such teachers are concerned with the task at hand, they take charge, they provide the leadership that is needed and they are prepared to direct the learning process, whatever method is used.

THE FIRST MEETING

The successful teacher of adults will conduct his first session with directives to himself, such as the following:

1. Get the advisory committee to work.
   - 1.1 Have each member bring three others to the meeting.
   - 1.2 Have the committee take care of the food and recreation (if these are desired).
   - 1.3 Make the committee chairman responsible for items of business, comments on next meeting, plan to secure additional enrollees.
2. Plan the lesson thoroughly.
3. Start on time.
4. Go through with the lesson as planned with your best effort (even if only a few members are present).
5. Secure evidence to evaluate your teaching.
6. Lead up to the next lesson.
7. Stop on time.

8. Have a business session if one is needed.
9. Have a recreation session.
10. Get acquainted with the members of the class.
11. Adjourn.

INTRODUCING THE LESSON

Teaching on the basis of a problem such as *What Variety of Soybeans Should Be Selected for Grain Production?,* the teacher should relate the problem to the needs of the class members so that each member recognizes its importance. This part of the teaching is to solve the first problem: Why should the students be involved with the problem to be discussed? The second part of the teaching is the actual solution of the problem. The first problem, in some situations, requires little attention because the students recognize the problem, and they have a felt need to solve it. In other situations, the predicted needs are not recognized and the teacher has the task of developing an understanding of the problem and its implications to the students. Often, making students recognize that they have a problem requires more teaching time and effort than actually solving the problem. For example, convincing a farmer that he needs to add lime to grow alfalfa could be more difficult than teaching him how to fulfill the lime requirements. This latter decision could be reached by calling the dealer and telling him what he wants, and when he wants it applied.

The student may be made to recognize the problem in many ways, but the process, the meaning and the relevance of the problem must be defined. For instance, when teaching how to select a variety of soybeans, the teacher must point out that the student's present variety is inadequate and that there are superior varieties. He can accomplish this by using available experimental data, showing pictures to compare the good with the less desirable, and referring to individuals who have had favorable results after making a change. He can take the students to visit a farmer who used a superior variety, allowing them to judge the results for themselves by observing the field. The teacher must exercise his own judgment in deciding how to secure student acceptance, but in the end, the students should be eager to learn how to solve their problems, and to put their decisions into practice. In solving such a problem as *Growing Soybeans,* the lecture method is not effective by itself; the students must be involved through discussion and understanding of the real situation. Student acceptance must be based upon fact, without over-selling and exaggeration. The teacher must present both the disadvantages and the advantages. He should be interested only in the truth and in getting better practices adopted.

REVIEWING THE INTRODUCTION

After the lesson has been introduced, the teacher may need to refer to some of the basic reasons for solving the problem. This is often necessary when the problem requires more than one session to solve, or when students feel that the effort required to make a change may be greater than the return. At these stages, they may need to be reconvinced and the teacher must secure a reaffirmation before he can successfully reach the objectives of the lesson.

SEGMENTING THE SUBJECT

The science of teaching points to the importance of systematic instruction and logical presentation of factual materials. Students should recognize what is to be taught and understand the analysis necessary to secure the desired results. The teaching procedure, as it moves from step to step, must make sense to the student.

In the title of this section, the word *subject* was used because subject can connote either a problem or a topic. There is some difference in segmenting a problem and in segmenting a topic. Segmenting a problem is separating it into minor problems with direct intention to meet the students' needs. Segmenting a topic is separating into parts solely on the basis of the material, with no concern for the students' personal needs. In the problem, the personal application is usually made manifest in the statement. In the topic, the teacher has to incorporate the personal application to students during the teaching procedure. Whether he uses the problem or the topic approach, his skill in getting acceptance to learn and in securing worthwhile outcomes determines his competence as a teacher.

SEGMENTING A PROBLEM

Taking the personal problem *How Should Harry Meet the Calcium Requirements for Growing Alfalfa?*, the minor problems or questions could be:

1. What is the calcium requirement for alfalfa?
2. What is the calcium content of the soil in the field set aside for growing alfalfa?
   - 2.1 What method should be used to take soil samples?
   - 2.2 Where should the samples be taken in the field?
   - 2.3 What procedure should be used to prepare the samples for testing?
   - 2.4 What procedure should be followed in testing for calcium content or securing a testing of the samples?
   - 2.5 What should be the interpretation of the tests?
3. When should calcium be applied on a field for growing alfalfa?
4. What material should be used?
5. What amounts should be applied?
6. What method of application should be used?

SEGMENTING A TOPIC

In a topic such as *Meeting Calcium Requirements for Alfalfa,* the segments could be:

1. Alfalfa requirements
2. Calcium carriers
   - 2.1 Characteristics
   - 2.2 Equivalents
   - 2.3 Costs
3. Soil testing
   - 3.1 Taking samples
   - 3.2 Preparing samples
   - 3.3 Testing for calcium or securing tests
   - 3.4 Interpreting tests
4. Applying the calcium carrier
   - 4.1 Amount
   - 4.2 Time
   - 4.3 Procedure

DETERMINING THE PROBABLE ANSWERS

One of the reasons for the accepted use of problem solving in teaching is that the procedure involved is consistent with the process of thinking. Thinking and problem solving are mind and body reactions that individuals make to satisfy felt needs. In recognizing and solving problems, the individual seeks a possible answer. He may find one or more good probable answers to test so he can arrive at a wise decision. He soon learns that the problem solving process is made more effective, and is less time consuming, when he includes only the most likely probable answers to explore. In teaching, the teacher should tactfully direct the selection of probable answers for the problem under consideration.

Number of Probable Answers Varies. Some problems may have only one probable answer, and others may have two, three, or more. In the problem of determining the calcium requirement of alfalfa, experimental tests show that, for best results, a certain pH is required. Here there is only one possible answer. This could likewise be true with such problems as, *At What Depth Should the Foundation Be Laid for the Hog Barn, When Should Spring Oats Be Seeded, What Procedures Should Be Followed in Taking Soil Samples, What Should Be the Cement Mix for a Concrete Driveway,* and *What Amount of Soybean Oil Meat Should Be Added to an Otherwise Established Feed Mixture to Get the Crude Protein Content to 16 Per Cent.*

Some problems have only two alternative answers, which are usually yes or no. Examples are: *Should the Hogs Be Sold Now; Should the Tracto Be Bought;* and *Should Additional Land Be Rented.* Some problems such as the following resolve themselves into two good probable answers;

(a) the selection of alfalfa seed from two apparently good and acceptable varieties, (b) the building of a poultry house for either caged or open pen housing, and (c) the use of the narrow or wide row in planting corn. It should be noted that the latter problems could also be stated on a yes or no basis. As an example, to the question, should a certain variety of alfalfa be used for seed, the answer would obviously be either yes or no.

The problems with three or more probable answers are common in life, and logically also in teaching. In the examples of segmenting a problem, the subsidiary problems when should calcium be applied, and what material should be used, have three or more probable answers. Taking the problem *When Should Harry Apply Calcium,* Harry could make the application at many times, but some of the most probable times are: (1) In the summer after the seedbed is prepared, and just before seeding the alfalfa; (2) In the winter on the crop preceding alfalfa; and (3) In the summer just before preparing the seedbed for alfalfa.

Limit the Number of Probable Answers. In the above problem, only three possible answers are presented, but obviously there could be many more. Successful teachers carefully screen and limit probable answers so that only the most likely are included. This minimizes the time required to solve problems, and keeps the teaching procedure practical and similar to the way intelligent people arrive at wise decisions.

Regardless of the problem, the probable answer or answers need to be tested. This requires the selection of criteria, factors or things to consider and then applying the best data available for decision making.

ARRIVING AT DECISIONS WITH A TOPICAL OUTLINE

In the case of the topic *Meeting the Calcium Requirement for Alfalfa,* the segmentation would include the important items to consider such as characteristics, equivalents, and costs, under Calcium Carriers. Facts are required to understand the characteristics of the calcium carriers, the amounts of each carrier needed to supply the calcium required, and the relative costs of calcium in each of the carriers. In good teaching, the teacher will help the members of the class to secure data relevant to their particular selections. Then it would be wise for teachers to check the adaptations that they make with the facts to arrive at specific applicable decisions. When this is done, decisions are made not unlike those discussed in teaching with a problem. Thus, the important thing to remember in teaching is not particularly the method used, other things being equal, but the attainment of objectives.

Securing Facts

SELECTING FACTS

Excellent materials are now available to teachers of agriculture from agencies such as State Offices of the Cooperative Extension Service, State

Agricultural Research and Development Centers, and State Departments of Education. By planning the program of instruction for an adult course well in advance of the first meeting, teachers have ample time to supplement their material inventories for almost every item included in the curriculum. Probably the greatest difficulty is in getting data that are adaptable to local situations. Fortunately for the resourceful teacher, time, when available, will help him find a way whether by study, test plots, observation, or inquiry.

The teacher of adults must secure appropriate data so wise decisions are made. His purpose is to help students to use these decisions so they can make desirable changes in practice. Often these changes require considerable time and money. These may not be readily available and the change consequently may be difficult to accomplish. The influencing factor could well be the confidence that the student has in the teacher. Thus the teacher should recognize his professional responsibilities and conduct himself so that he merits the faith his students should have in him. Such a teacher will be concerned to continue his assistance so that change becomes a success.

UNDERSTANDING LOCAL SITUATIONS

A frequent criticism of teachers by adult farmers is that they teach from the book. What the farmers are saying, in effect, is that teachers do not have the facts relevant to their situations and so cannot translate the technical information into practice. An obvious reason could be that the teachers have not been on the farms or in the business of students and so do not know the real problems that they must solve. Such teachers are of little help to their students in adopting the subject so it can be used. They fall short of meeting the meaningful teaching objectives of improving practice and changing behavior.

Teachers who are successful make every effort to know their students. They talk to them regarding problems and situations before and after class, on chance meetings, and on their farms or places of business. Here they can learn first hand what to do to help their students.

USING THE RESOURCES OF THE CLASS

Adults may have some disadvantages as learners, but one of their advantages over high school students is their experience and accrued intelligence. Combined intelligence is not only in depth but covers many fields. Successful teachers have learned to utilize this body of knowledge. It not only is a readily available resource of information, but provides the opportunity for each member to make his unique contribution in the teaching-learning situation. Consequently two aids in learning are achieved—useful information and student involvement. These are made possible by such means as effective questioning, presentations, panels, symposiums, forums, role playing, and socio-dramas.

ELICITING INFORMATION BY QUESTIONING

Questioning should be done so that the teacher gets a response in keeping with the intent of the question. Teachers may help themselves by formulating their questions when they plan their lessons. They can anticipate the questions that will secure the responses they want. These questions may be aimed at clarifying situations, checking probable situations, securing information, evaluating data, formulating decisions, and adapting decisions to situations (3, pp. 113-19).

> Throughout the course of educational history the question has been one of the most common, if not the most common of teaching techniques. It continues to be so in spite of modern changes in educational theory, for it is a test, for checking memory and understanding, and getting at higher learning. It has many uses in the classroom. Among these uses we find the following mentioned in textbooks on teaching:
>
> 1. To find out something one did not know.
> 2. To find out whether someone knows something.
> 3. To develop the ability to think.
> 4. To motivate pupil learning.
> 5. To provide drill or practice.
> 6. To help pupils organize materials.
> 7. To help pupils interpret materials.
> 8. To emphasize important points.
> 9. To show relationships, such as cause and effect.
> 10. To discover pupil interests.
> 11. To develop appreciation.
> 12. To provide review.
> 13. To give practice in expression.
> 14. To reveal mental processes.
> 15. To show agreement or disagreement.
> 16. To establish rapport with pupils.
> 17. To diagnose.
> 18. To evaluate.
> 19. To obtain the attention of wandering minds.

Good questions stimulate class participation (8, p. 200), which encourages discussion among students with the teacher directing the procedure. It reduces the teacher-to-student-to-teacher dialogue, which could be uninteresting if pursued continuously. Teachers who secure class participation have time to think and to plan the next steps in developing the lesson. It puts students on the offensive to commit their thinking and then on the defensive to justify their statements. The teacher should not be in one side of a debate with the class, but instead should be facilitating learning through student participation. This latter fact cannot be ignored, because students do attend classes to also learn from their peers.

Observing a teacher who can secure controlled interaction among students, and in so doing stimulate interest in class proceedings, is a thrilling experience. It seems that this teaching is so effortless and yet so constructive. Certainly there is an art to acquire in questioning. Probably a recognition of the elements common in good questioning would be helpful. Then if these elements were incorporated in the questions asked, the effect could be determined. By continued practice with these elements, improvement in the art would make teaching easier and most certainly more rewarding.

Qualities of Good Questions. The aforementioned attributes of good questions enumerated in statements are:

1. They are clear, definite and concise.
2. They start with why, what, which, etc.
3. They are interesting and timely.
4. They are thought provoking.
5. They are adapted to individual differences.
6. They encourage expression and discourage guessing.
7. They invite questions.
8. They stimulate effort.

How would you rate the following questions?

1. How about John's calcium needs?
2. John, how do you feel about this type of calcium?
3. Are you interested in improving soil fertility?
4. Recognizing the differences of soils, the probability of error in soils sampling, the effect of soil and weather conditions, would you give the range in amount (maximums and minimums) that should be used to satisfy the needs of a 100-bushel corn crop as related to the application of calcium in any of its commercial forms, considering, of course, that soil type affects quantities of calcium to properly compensate for the variableness of buffer action?
5. How do the overall economic conditions effect the cost, use, and profitability of calcium applications?
6. What facts support the decision to use calcium carbonate rather than calcium hydroxide for correcting soil acidity?
7. Which of the liming procedures outlined on the chalkboard requires the larger initial capital outlay?
8. Which will give the greater net return for the next five years?
9. What difference does it make whether John uses calcium carbonate or calcium oxide to get a neutral soil reaction?

10. Which crops grown in this community are most sensitive to calcium deficiencies?
11. What determines the reliability of a soil test?

Using Individual Study

CLASS STUDY

Another way to secure additional facts and to verify the appriopriateness of those presented is through individual study. Adults have a limited time in class, so their time must be used efficiently. Some studying in class will be accepted, if the material supplies a ready source of data, if it is well selected, if it is readable, and if it does not cause embarrassment to those who do not read well.

HOME STUDY

Probably the best way for students to secure information and develop understanding is through home study. Most adults will read, if they have appropriate materials. Some motivation may need to be developed by tactfully complimenting those who pursue individual study and by using the knowledge that they have acquired. The real problem for the teacher is primarily that of securing study materials. He should exploit the related information in farm magazines, secure sufficient copies of bulletins that apply and can be taken home from various agencies, prepare mimeographs that are applicable, and cite materials that can be bought. Teachers taking this extra step will find the results rewarding. Once this studying outside of class is initiated by a few, it spreads to others. Eventually many of the adults will develop good reference libraries on their own. They will learn how to help themselves. This, when accomplished by a teacher, is teaching at its best.

ORGANIZING FACTS

Facts must be organized so that they may serve their purpose. When using the problem approach, facts needed should be obviously associated with the specifics of the problem. Likewise when using the subject approach, the analyzing process should suggest the information that is needed to develop an understanding of the subject.

Coming back to the problem of *What Variety of Soybeans Should John Select for Grain Production,* the objective is quite obvious, but organizing the facts for decision making may be somewhat difficult. When we determine things to consider in solving the problem, such as the number of days from seeding to maturity, resistance to root rot, resistance to lodging and test plot yields we have little or no difficulty in categorizing the information. When a form similar to the following is presented, the students can easily recognize where facts are needed to arrive at intelligent decisions.

What Variety of Soybeans Should John Select for Grain Production

Items to Consider	Possible Varieties			
	A	B	C	D
Number of days from seeding to maturity				
Resistance to Root Rot				
Resistance to Lodging				
Test Yields				

Arriving At Decisions

EVALUATING THE FACTS

Teaching students to use facts properly in making intelligent decisions is important. Upon these decisions hinge the practices that are adopted and the degree to which the lesson objectives are achieved. Students should recognize the factors involved or criteria considered and their relative importance to the problem. Then as they evaluate the facts relevant to each factor, their decisions should be indicative of intelligence. Obviously, in evaluations, items considered are not given similar values by students so it is not likely that they will arrive at the same decisions. This may in itself be good, because it permits a reviewing of the evaluation of the different decisions and may help students to develop a better understanding of the problem and its solution. Taking the problem of *What Soybean Variety Should John Select for Grain Production,* the importance given to lodging resistance as compared with the other factors will affect the decisions. There may be no one right answer for the class, but John, who may have a problem with lodging soybeans, should give emphasis to varieties resistant to the characteristic. Harry may be concerned with eliminating root rot, so his decision will be weighed to resistant varieties.

FOLLOWING THROUGH DECISION MAKING

A common mistake made by teachers in problem solving is to stop after they have solved a personal or general problem and assume that the students in the class can take the decisions from the personal or general problem and apply them to their individual situations. Time should be taken in class for students to apply the facts to their conditions and to get class reaction to their decisions. It is through this process that facts are used and the complete act of thinking is effected. This also provides an excellent learning opportunity to show how situations have variables and how variables affect decisions. Students who can make wise decisions using facts in different situations are being taught properly. The intellectualizing that takes place when facts are applied to situations is indicative of understanding the facts and is a first and necessary step in getting the adoption of practices.

The teacher may purposefully stop the discussion after allowing a reasonable time for the application of facts to individual decisions and then move on to the next problem or subject. Hopefully he will do this, knowing that after class, on the farm, or at some other time, he will not forget to give additional help to those who need it.

ALLOWING FOR INDIVIDUAL DIFFERENCES

From the foregoing discussion, it should not be inferred that mature adults are intellectually incompetent to do things for themselves. Individuals in the class differ. Some are quite ingenious and can initiate plans with no additional help. Some are hesitant and cautious, particularly when capital investment is involved, and need the stimulating help of a leader to adopt the practices. The successful teacher knows his students. He takes class time to discuss why the problem or topic is important, and how the problem can be solved. He knows he is teaching individuals in a class and will go beyond class time when needed to be certain that the objectives of the lesson can be attained.

TEACHER RESPONSIBILITY IN DECISION MAKING

The concluding decision for the problem should show good thinking as evidence of good teaching. It should result from the proper selection and use of facts and the efficient use of time. It should be the result of student involvement and be consistent with the teacher's plan for the lesson. This does not mean that the teacher and the class must always agree or that either cannot be wrong. If the teacher is in error, he should recognize where and why he made the mistake and change his plans in teaching accordingly. If the students are in error, the teacher should use tact in reviewing the facts and their evaluation so as to eliminate the mistake made.

The teacher should be alert to the possibility of meeting the challenge of a few students who willfully or otherwise may unduly influence the thinking of the group. Generally adults know their peers quite well and are not easily misled. They also respond eagerly when competent teachers question class decisions.

PLANNING THE APPLICATION OF DECISIONS

After class discussion and evaluation of facts, the decisions made in problem solving are relevant to the case situation used. If the teaching has been conducted on a subject outline, the conclusions are probably general. These conclusions, like the decisions from problem solving, must be tailored to fit the situations of the individuals in the class. It was suggested earlier that after decisions are made in the class, the teacher should give students time to intellectualize and make decisions appropriate to their individual situations. If students need additional help, this should be provided.

This step in teaching has several justifications. One is that if the objectives of teaching are achieved, the individuals improve their practices and change their behavior. Thus, general conclusions or decisions have no merit in themselves, but only when they are transformed into action by individuals. The extent to which these changes are made, of course, determines the quality of the instruction. A second reason for planning the application of decisions is that there is often much more involved in getting a practice adopted than making a decision. Here is where class instruction gives way to individual initiative and planning. Class time is usually not available for students to completely plan the application of decisions.

INDIVIDUALIZING PLANS

The teacher can assist students by using class time to take one or two students through the process of developing plans for adapting decisions for individual use. Using as an example the selection of a variety of soybeans, John may not have the necessary capital available to permit him to use certified seed. His equipment may limit his choice of seeding to drilling. His soil, his rotation, and his rental agreement could also affect the application of his decision. Helping John in the planning could very well bring about a change from inaction to action. Since improving practice is an objective of teaching, stopping short of the goal can not be justified. Developing individual plans for action is another important step in teaching adults. It should be part of teaching in the classroom and is often neglected.

Including Participating Experience

IMPORTANCE OF PARTICIPATION EXPERIENCE

Involvement of the learner in the teaching-learning situation is essential to learning. This is true whether the objectives of the lesson require mental or physical skills, or a combination of both. We have discussed how to secure student involvement in making managerial decisions, but so far nothing has been discussed regarding student participation in a class situation where manipulative skills are to be taught.

IMPORTANCE OF MANIPULATIVE SKILL

Teachers frequently neglect to teach manipulative skills to adults. The reason may be that the teacher does not have the extra time required to provide the laboratory experience, school facilities are inadequate, needed field trips are difficult to plan and conduct, the teacher feels inadequate to conduct demonstrations, or the teacher does not recognize a need to teach manipulative skills. But no reason can be justified when the lesson objectives require the teaching of manipulative skills. Teachers of adults should account for all the vocational needs of adults and leave no neglected areas that interfere with the adoption of improved practices. When instructional programs are complete, the teaching of some manipulative skills seems imperative. Where they are included, the lesson plan should be made to provide participating experiences for the students.

PLANNING THE EXPERIENCE

In solving the problem, *Meeting the Calcium Requirements for Growing Alfalfa,* some manipulative skills are required. The outline that follows suggests a procedure to teach students some of the essential skills and understandings on how to properly take soil samples:

How to Take Soil Samples

1. Show the importance of accurate sampling.
   - 1.1 What would be the result if too little calcium were applied?
   - 1.2 What would be the result if too much calcium were applied?
   - 1.3 What will happen when some sections of the field need more calcium than others, and this difference is not accurately determined by the test?
   - 1.4 What tests really determine the amount of calcium that is needed—the high tests, the low tests, or the mean of the tests?

2. Show where the samples are to be taken.
   2.1 What determines where samples are to be taken?
   2.2 What number of samples should be taken?
   2.3 What number of samples may be used in a composite sample?
3. Acquaint them with the tools that are needed to take the sample.
   3.1 The augur
   3.2 The spade
   3.3 The containers
4. Take the class to the field to demonstrate how to take samples and then have them take samples.
   4.1 Demonstrate what is done when taking soil samples.
   4.2 Have each participant take samples.
   4.3 Observe the students in taking the samples.
   4.4 Demonstrate again those parts that were not done properly.
   4.5 Answer the questions raised by students.
5. Return to the classroom to complete the lesson. Consider such things as labeling, drying, and packaging of the samples.

GETTING ADULTS TO PARTICIPATE

Adult students are sometimes reluctant to participate as individuals in the presence of a group of peers. This reluctance decreases as individuals become acclimated to the teacher and to their classmates. Having frequent lessons where demonstrations with adult laboratory experiences are involved tends to make adults less reserved and more at ease. Being careful in evaluation and trying not to show awareness that adults are shy when working with the class will help. Having the adults work individually or with their own special group will aid in overcoming this oversensitivity to what others may think of their performance.

PARTICIPATING EXPERIENCE OUTSIDE OF CLASS

Many of the skills needed to adopt a practice can be acquired by the individual using the class instruction and his own ingenuity and initiative. A demonstration in class on how to select planter plates for the various sizes of seeds should be sufficient for a student to acquire the skill on his own. In the same category would be teaching how to use common farm tools, where practices at home would develop the needed skills. Taking time to secure practices in performing skills is not unlike taking time to study to acquire knowledge. Both are to be encouraged. They result from effective teaching and interested students.

DETERMINING THE AIDS TO USE

Aids should be as the word denotes, aids to teaching. They should add something to the situation. They should utilize to a greater degree, either singly or in composite, the senses of perception. In effect, aids should expedite the teaching-learning process and make teaching and learning easier. The effectiveness of teaching aids is related to the teacher's competence in their use. This is highly personal, and each teacher in perfecting his teaching performance will have to make such decisions for himself. Certainly he can get help from good teachers. They can observe him teach and he can observe them.

Some State Departments of Agricultural Education either have developed, or have access to, materials to assist teachers of adults. This is likewise true for some State Offices of the Cooperative Extension Service. The aids in many situations relate directly to the subjects commonly taught to adult groups. The teacher's task is to properly screen from the relevant materials that which will make a contribution to the teaching.

Certainly he will never reach the point in his teaching skill where his selection and use of aids will be perfect. However, he can become increasingly effective as long as he continues his efforts to improve. It is such individuals that we call Master Teachers.

The teachers that we admire most are outstanding because of certain attributes. One may have been a master in the use of the chalkboard. He may have been able to write legibly. The materials written may have been well organized. He may have been able to use the chalkboard to vividly illustrate what otherwise would have been difficult to comprehend. Another good teacher may have been able to use visuals appropriately. He probably was able to correlate each visual with aspects of the lesson that allowed no dull moment. Another teacher may have been a master because he was able to arouse interest in individual study of related references, in using community resources, and in conducting demonstrations. Then there were a few like Mark Hopkins who had the gift of speaking and of questioning with such appropriateness that if they incorporated anything additional, it would have been distracting. One could conclude that there are many aids and many ways to properly use them with no convenient prescription to guarantee success.

REFERENCES

1. Bryson, Lyman, *Adult Education.* New York: American Book Company, 1933.
2. ________, *Adult Education.* New York: American Book Company, 1936.
3. Clark, Leonard and Irving S. Starr, *Secondary School Teaching Methods.* New York: The Macmillan Company, 1959.

4. Crow, Lester D. and Alice Crow, *The Student Teacher in the Secondary School.* New York: David McKay Co., Inc., 1964.
5. Devor, John W., *The Experience of Student Teaching.* New York: The Macmillan Company, 1964.
6. Hammonds, Carsie and Carl F. Lamar, *Teaching Vocations.* The Interstate Printers & Publishers, Inc., 1968.
7. Sheats, Paul H., *et al., Adult Education.* New York: The Dryden Press, 1953.
8. Woodruff, Asthel D., *Basic Concepts of Teaching.* San Francisco: Chandler Publishing Company, 1962.

ADDITIONAL REFERENCES

Bergevin, Paul Emil, *A Philosophy for Adult Education.* New York: Seabury Press, 1967.

Bergevin, Paul Emil, and John McKinley, *Participation Training for Adult Education.* St. Louis: The Bethany Press, 1969.

Cotton, Webster E., *On Behalf of Adult Education.* Brookline, Mass.: Center for Study of Liberal Education for Adults at Boston University, 1968.

Hollenbeck, Wilbur C., *et al., Community and Adult Education.* Chicago: Adult Education Association of the U.S.A., 1967.

Jensen, Gale, *et al., Adult Education: Outline of an Emerging Field of University Study.* Chicago: Adult Education Association of the U.S.A., 1964.

Miller, Harry L., *Teaching and Learning in Adult Education.* New York: The Macmillan Company, 1964.

Morgan, Barton, *et al., Methods in Adult Education.* Danville, Ill.: The Interstate Printers & Publishers, Inc., 1963.

Sattler, William M. and N. Edd Miller, *Discussion and Conference,* (2nd ed.). Englewood Cliffs, N. J.: Prentice-Hall, Inc., 1968.

Chapter 7 SUPPLEMENTING THE TEACHING IN THE CLASSROOM

In the previous chapter, teaching in class situations was the major consideration. Emphasis was given on how to involve students in discussion, individual study, planning and adopting practices. Good teachers can also involve students individually or in small groups to effect behavioral changes. In addition, they can use the teaching opportunities provided by field trips, resource personnel, tests, demonstration, and records. Admittedly there are many others that could be added and the aspiring professional should not limit his repertoire of aids in teaching adults to those presented in this chapter (4).

A descriptive account of the use of individual instruction, small group instruction, field trips, resource personnel, tests, demonstrations, and records could be quite an extensive volume in itself. In an effort to keep explanations to a minimum and not omit significant steps, outlines are frequently used in this chapter.

Small Group Instruction

Often teachers will have situations where a group of students within a class will have similar needs and could benefit from additional teaching assistance. For example, a group of farmers with an interest in swine may want help in designing farrowing quarters or treating pigs for worms. The teacher can arrange a special class for these students. It could be at one of their home farms, in a shop, or wherever it would be best to do the teaching. There is nothing particularly different about this kind of instruction than that used with a larger group of students. The suggestions offered in the previous chapter should be helpful in planning the teaching of a small class. However, there can be less formality and certainly more student participation and teacher attention to individual needs when fewer students are involved.

Such teaching, when provided, can expedite the regular class sessions. It reduces the time spent in class in helping the few students where special needs may not be relevant to many of the others in the class. Some teachers use the technique frequently and with excellent results. It certainly has a place in teaching adults when conditions warrant. This instruction should not be confused with group instruction, in which a class is divided into committees that work on special assignments for the achievement of common goals. Such group instruction involves the teacher with concurrently functioning groups.

The using of simultaneously functioning groups, often called "buzz sessions," has much to offer a teacher who is able to use the technique effectively. The experience it provides for adults will help them in life situations. Here individuals need to rely upon their own initiative for the solution of problems. It also helps them to work together for common objectives, to present and defend points of view, and to be understanding, intelligent citizens in a democracy.

In the following several pages some suggestions are provided for teachers of adults in the use of small group instruction. In the presentation, attention is given to reasons for such instruction, attributes of teachers, indicators of student abilities, reasons for group failures, recognizing student sensitivities, kinds of discussions suitable, directives to teachers, formulating discussion questions, and indicators of group maturity.

I. Some reasons for the use of small group instruction:
   A. It gives each individual more opportunity to participate and enhances his feeling of belonging, thus preparing him to better fill his place in a democratic society.
   B. It provides more interaction among individuals—each member stimulating the others.
   C. It provides more student-centered and less teacher-centered activity.
   D. It provides a good procedure for securing solutions to controversial questions that may arise.

E. It results in decisions made by students in a group of their peers which are more meaningful and which are more likely to be abided by than decisions made under the influence of a teacher.
F. It provides an activity for short intervals, adds variety to classroom procedure, and provides a "break" from the monotony of regular routine.
G. It results in time saving after teacher and students have acquired skills in group activity.
H. It provides practice in policy and program development which leads to better self-government.
I. It provides an excellent learning experience and more than a memory routine. (It is give and take in a life-like situation.)
J. It gives students experience in an important democratic process since more of all work today is group activity and we need to learn the ways of effective group practice.

II. Some attributes of teachers able to utilitze groups successfully:
A. "Feel" sensitivity of the group.
B. Respect the personality of the members.
C. Respect the freedom of an individual member.
D. Provide the opportunity for all to participate in a decision that involves themselves.
E. Realize that a leader is best when the members barely know that he leads.
F. Have and keep a united group.
G. Do not become discouraged—groups must be taught how to operate.
H. Always keep an open mind.
I. Appreciate scientific method of problem solving.
J. Appreciate and like the problem-centered approach.
K. Derive principles out of experience, rather than compel adjustment to so-called principles.

III. Indicators of student abilities to assume group activities:
A. Students need to make progress in working together in a spirit of tolerance, cooperation, honesty, and fair play in order to proceed from the simple to the more difficult stages of mature group activity. Maturity in group activity means that individuals have the ability to:
1. Distinguish between facts and opinions
2. Distinguish between information and judgment
3. Distinguish between leadership and dictation

4. Distinguish between democracy and dominance
5. Distinguish between quibbling and contribution
6. Distinguish between debate and problem-solving
7. Distinguish between a point of view and prejudice

B. Stage I Teacher should start group activity by using the small group "huddle."

C. Stage II Fact-finding may be the next stage in training for group activity. Training in reference work —using the resource materials in the vocational agriculture department—is a requirement for success.

D. Stage III Training in evaluating information—being being able to distinguish the important from the less important information and knowing where and how it applies.

E. Stage IV Making decisions on specific problems and applying factors and facts to solving problems of other individuals in the group. This approaches the goal of each individual making right decisions on his own problems without teacher-direction.

F. Stage V The group finally progresses to the maturity stage when it is able to analyze problems in the light of specific situations.

IV. Some reasons why groups fail:

A. Early training did not encourage discussion
1. At home (parental domination)
2. At school (answer teacher's questions)
3. At church (listen)
4. Generally enjoying listening
5. Generally depends upon others
6. Individual winner vs. group welfare

B. Physical environment of meeting not good
1. Room too large
2. No physical separation
3. Poor seating arrangement for comfort or friendliness
4. Room stuffy, cold, hot, poorly lighted

C. Unsuccessful in preparing group for discussion
1. No feeling of belonging—students feel isolated, not acquainted
2. Poor choice of subject
   a "No interest in it"
   b "We don't know enough about it"
   c "Discussing this subject doesn't make sense"
   d "We can't do anything about it"

3. Topic poorly handled
    a "Teacher tells all—nothing left to discuss"
    b "Incomplete introduction"
    c "Not challenging"
4. Questions poorly worded
5. Feeling of futility
6. Participants not involved in decision-making
7. Old authoritarian classroom atmosphere stifles group activity

D. Domination by others
1. Leader who does not
    a Give opportunity
    b Give encouragement
    c Have necessary skills
2. Expert who does not play the role in a group and creates passivity or dependence
3. Monopolizing member who does not know the role of a good member.
4. The "brass" who fail to realize that few in their presence would "stick their necks out."

E. Fear of ridicule
1. By friends
2. Fear of deficiency in
    a Grammar
    b Expression
    c Emotional control—blushing
    d Fluency
3. Fear of appearing stupid

F. General feeling of inferiority
1. "I don't know much—let the others talk"
2. Don't dare talk
    a In such a large place
    b In front of people
3. "I can't say what I mean when I am on my feet"
4. Fear of being contradicted and of inability to "hold my own" in an argument
5. Fear of offending others
6. Fear of expressing what may be a minority opinion

V. Some characteristics of students that teachers should recognize when using group methods for instruction:

A. Teachers must be conscious of individual thoughts of the students who will make up the group. This is a part of feeling the "sensitivity" of the group. If groups are to function properly, the teacher will keep the following points in mind because *group members say* "If my best efforts are wanted, someone will consider the fact that:

B. I need a sense of belonging
   1. A feeling that no one objects to my presence
   2. A feeling that I am sincerely welcome
   3. A feeling that I am honestly needed for myself, not for my hands, my money, etc.

C. I need to have a share in planning the group goals.

D. I need to feel that the goals are within reach and that they make sense to me.

E. I need to feel that what I'm doing does contribute to the welfare of people—that it extends in purpose beyond the group itself.

F. I need to share in making the rules of the group—the rules by which together we shall live and work toward our goals.

G. I need to know in some clear detail just what is expected of me so that I can work confidently.

H. I need to have responsibilities that challenge, that are within range of my abilities, and that contribute toward reaching our goals.

I. I need to see progress is being made toward the goal we have set.

J. I need to have confidence in our teacher. This confidence will be based finally upon my assurance of consistent fair treatment from the teacher, the recognition when it is due, and trust that loyalty will bring an increased measure of security.

K. I need, at any given time, to conclude, "This situation makes sense to me."

L. I need to feel that this is learning, progress, fun.

VI. Examples of kinds of discussion suitable for group instruction:

A. *Policy-making discussion.* Example: "What shall be the policy of the state regulatory agencies in the use of pesticides?"

B. *Policy-evaluating discussion.* Example: "In what one way might we best improve the Land Diversion Program?"

C. *Program-building discussions.* Example: Each committee of the adult class has submitted its suggestions for improving the net return of the corn enterprise. "What one addition would make these reports more beneficial to all of us?"

D. *Fact-finding discussions—decisions.* Example: Problem-solving in classes. Steps in problem-solving with groups:

1. Discuss problem with entire class
2. Break into committees and elect a chairman and secretary
3. Search for facts
4. Determine possible solutions
5. Select one best solution
6. Report to entire class

E. *Coordinating discussions.* Example: Elect a committee from the group to review and prepare a summary of the procedures for the reports of each group on improving the net return from the corn enterprise.

F. *Decision—action discussions.* Example: Who should be selected to develop a program of instruction for our adult class.

VII. A teacher's brief for group operation:

A. Be sure physical environment offers the right atmosphere.

B. Be sure members are psychologically ready for discussion—prepared by teacher.

C. Get entire group into committees.

D. If unacquainted encourage talking by giving names, profession, etc. (One to two minutes)

E. Each group select a chairman and a secretary-spokesman. (One-half to one minute)

F. The secretary is asked to keep a record of the discussion by:
1. Writing the question at the top of the card
2. Writing all answers given
3. Not writing names

G. The chairman is responsible for his group, and he should have:
1. The secretary write the question on the card
2. The secretary read the question before there are any answers
3. Each member screen out best possible solution for one-half to one minute before any discussion is allowed. (Don't put down individual answers and add them up. Better, let questions start at once on a "Let's see. Is this what the question at hand means?" basis.)
4. Each person state best possible answer before there is any discussion.

H. Ask the question or state the problem: "You will be allowed five minutes to record and discuss your statements."

I. "Mr. Chairman, remember your instructions?" "Are there any *questions*?" "O.K.! Let's get started."
J. The leader should move among the group to see if there need be further explanation. *Do not* stay near one group.
K. After appropriate time, determine the approximate length of time necessary to finish the assignment.
L. Give a warning: "Mr. Chairman, you have just one minute to make certain all have given an answer."
M. "Mr. Chairman, you have two minutes to screen out the best idea. Mr. Secretary, circle or check the best idea."
N. You are ready for reports that best suit the situation. It might be:
   1. An oral report by the secretary on the findings, conclusions, or decision.
   2. A demonstration by the entire small group.
   3. A collection of papers for use by a committee who will report back to the entire class.
   4. It may be necessary to go back into groups to pick the one best answer from the cards.
   5. In any case, collect the cards and indicate the use to which they will be put.
O. Always put on pressure by breaking into the discussion, and keep putting on pressure by time limits.
P. The time limits are only suggested and will vary according to the type of group discussion.

VIII. Some criteria for formulating good discussion questions:
A. Are the questions singular?
B. Will the questions be constructive to the group or individuals?
C. Are the questions involving so that thinking will be encouraged?
D. Are attainable goals involved in the question?
E. Are the questions selective?
F. Are the questions clearly stated?
G. Are the questions brief or concise?

IX. Some sample questions for specific purposes:
A. *To get acquainted or as an ice breaker.* Example: "What experience in your life was the most embarrassing?"
B. *To raise questions or ideas before or after a young farmer meeting with a resource person.*
   1. Before the meeting: "Mr. ________ is to be our speaker next week. His subject is ________. His qualifications and experience are ________. What particular phase or question in this field would interest you most?"

2. After the meeting: "What one question would you most like to ask in order to round out your information on this particular subject?"

C. *To canvass attitudes or to evaluate.*
   1. "In what one way do you feel that the demonstration on clipping dairy cows might be improved?"
   2. "In view of the discussion, what speaker should we get for the banquet?"

D. *To build agenda.* Example: "What one subject (or activity), if included in our future program, would be so meaningful to you that you would give up almost anything else in order to be present?"

E. *To exchange experiences.* Example: "What is the best idea that you have shared in or heard about that relates to our discussion this evening?"

F. *To nominate officers.* Example: "What two or three people in our class would you be most proud to call 'our president'?"

G. *To motivate thinking on a class assignment.* Example: "What in your opinion is the greatest implication of the recent presentation on crop rotations in relation to soil tilth?"

X. How teachers can determine when a group is mature:

A. A mature group is a self-directing, self-controlling body in which every member carries his part of the responsibilities for developing and executing the group's plans. The maturity of the group should be evaluated on the basis of:

B. The extent and quality of participation.

C. Its skill and facility in solving problems and its ability in staying on the problem.

D. Its understanding of and its ability to make adjustments in group processes.

E. Its intelligent management of its environment.

Field Trips

The term field trip may be a carry-over from the days when teachers took their classes to the field for study. In this day, the term has a broader context. It refers to class going away from the school, to such places as farms, homesteads, places of business, manufacturing plants, colleges, universities, and experimental farms.

Bergevin, Morris and Smith define the field trip as (1, p. 74):

A carefully planned educational tour in which a group visits an object or place of interest for first hand observation and study. The learning group makes its tour under the guidance of a person well-informed about the area of consideration. The person answers questions and points out features that might not be seen by the learners. The field trip should be followed by careful analysis, interpretation, or discussion of the places visited and the information obtained. A group can explain the significance of what they have observed on a field trip using other techniques, such as group discussion, a panel forum, or a speech forum.

The use of the field trip for adult classes may be less extensive now than in former years. Some of the reasons could be that the travel from school to the place for the trip may be too difficult because of congestion or distance. The time arranged for the trip may be an inconvenience or objectional to the students. The effort required to arrange a trip may be greater than the class wants to expend. The cost may be prohibitive. The weather may be unfavorable.

Even with the many plausible causes for the declining use of the field trip by some teachers, its merits certainly justify its consideration. Some of these are:

It keeps the teaching objective and realistic
It develops interest
It makes teaching easier
It utilizes local resources
It provides a reference for later discussions
It allows for direct observation
It provides a teaching situation in a natural setting

GETTING ADULTS TO RECOGNIZE NEEDS

Observing the practice under working conditions, having its merits discussed, and being able to ask questions of the operator provides a superior setting to motive learning. This is a teaching-learning situation at its best. Several years ago, a teacher of adult farmers had difficulty getting his students to consider an improved procedure in feeding dairy cattle during the summer months, until after they experienced a field trip. Here they observed what a successful farmer was doing that was entirely within their potential to do. Following the trip they immediately took the necessary steps to change their feeding programs. The results from the change in practice proved to be profitable. Getting farmers to want to change their practices is enhanced with good field trips. Here the teacher may lead the members of the class to observe the practices that they should be using. This often would be most difficult to do in the classroom through verbalization and without the seeing for believing.

Using the field trip to get acceptance of predicated needs should logically be timed to start a unit of instruction or shortly thereafter. Such trips that come early in the teaching of units are a ready source of reference during the teaching of the unit, and also for other units that follow.

DEVELOPING UNDERSTANDING AND SKILLS

After the lesson is underway, there are often aspects that are difficult to teach in the classroom, but are relatively easy during field trips. Previously it was shown how taking soil samples could be taught in the field. Other tasks that can be taught effectively in the field are the adjustment of mowers, combines, and balers, the installation of feed handling equipment and of grain drying equipment, and the planting and pruning of trees and shrubs.

GETTING ADOPTION OF PRACTICES

Often a field trip is useful to help the learner to make up his mind to adopt a practice. He may have the understanding and a workable plan to adopt a practice but yet is not quite ready to go ahead. Further discussion could have a negative effect. Taking the class on a field trip to visit several establishments that are using the practices can often get favorable results. Several incidents may be cited when the resourceful teacher used field trips to get some of the students "off-center" and activated. A trip in the early spring to a farmer with a paved barn lot motivated several in the class to make similar improvements. Likewise, a summer trip to farms where pastures had been improved resulted in many favorable changes in practices related to pasture management.

SPECIAL USES FOR TRIPS

Some teachers begin a lesson by an observation such as may be secured on a farm. After the teacher is satisfied that the visualization aspect has been fulfilled, the class may then return to a more desirable place to discuss the observation as it is related to the problem for consideration.

Since much of the teaching of adults in agricultural occupations occurs during the winter months, it is often difficult or impossible to correlate trips time-wise with class discussion. Many observational trips will, of necessity, be planned when there are things to see. The effect of cultural practices can best be studied during the growing season. Thus teachers should prepare their students for these learning experiences, conduct field trips, and then utilize what was observed later in the year as called for in the curriculum. It is probably not a great educational disadvantage for adults to make field trip observations that are not immediately associated with the finalizing of decisions. Certainly programs of instruction have to be prepared in late winter or early spring for the year following, so as to incorporate the appropriate field trips. Teachers following the procedure will likely find that the favorable conditions for the trips, including weather and work loads, correspond to a favorable attendance by adults. Of course, there will need to be attention given to publicity and such other preparation as one associates with a successful field trip.

Often teachers have the opportunity to plan tours or field trips with other agricultural agencies. These can be invaluable to students. They should be informed of the details and then given a little extra encouragement to participate.

PLANNING FIELD TRIPS

Much of the success of the field trip is in its planning. When facilities and personnel that are not directly under the control of the teacher are involved in conducting a field trip, pre-planning to insure proper coordination is a must. Consequently, arranging for all the essential details is no small task if the teacher hopes to effectively accomplish the maximum for his students.

IDENTIFYING THE FIELD TRIP

Develop a program of instruction of the adult course.
Determine the number of field trips that are needed to conduct the course.
Determine the kinds of field trips that are needed to conduct the course.
Determine where the field trips are to be taken.
Determine when the trips are to be taken.
Determine who is to make the arrangements for the trips.

A RECIPE FOR FIELD TRIPS

A. Determine exactly what is to be done on the trip.
B. Make the necessary arrangements with the host well in advance of the trip.
    1. Mention the size of the group.
    2. Agree on the arrival and leaving time.
    3. Get an agreement on what is to be done.
    4. Get an agreement on what the host is to do.
    5. Determine the limitations, peculiarities and precautions.
    6. Determine the special items to include.
C. Plan the teaching so that the field trip comes at the right time.
D. Prepare the class for the trip.
    1. Review the items of B that apply.
    2. Discuss plan to be followed.
    3. Discuss student participation.
    4. Discuss student responsibilities.

E. Arrange the transportation.
F. Arrange for paying the costs.
G. Conduct the trip as planned.
H. Thank the host.
I. Summarize the findings or the observation in keeping with the purpose of the trip.
J. Evaluate the experience.

Resource Personnel

Teachers of adults should constantly be aware of opportunities to improve the quality of instruction, and thus to enhance the learning opportunities of the students enrolled. When others can be involved to advantage and can be made available, they should be used. As was mentioned earlier, the members of the adult class can be sources of useful information when employed by the teacher. However, in this discussion, resource personnel will be considered the outsiders that are brought in to assist in the teaching. The use of resource personnel has at times been frowned upon by leaders in the field, and at times by the students. These negative reactions have, to an extent undeservedly, lowered the image of this aid in teaching. Really it deserves a better rating because resource personnel can fill a vital role in adult teaching. They can bring to the class new insights. They can break the monotony of having one teacher before the class, session after session. They offer something different. They can stimulate a class. But, they must be properly used.

Generally a good teacher who knows his students, their home situations and has worked with a class is preferred by the student when the teacher has the competence to teach the class. When this is the situation, the teacher should not use outsiders. The regular teacher is usually the best qualified to begin the session, to conclude the session, and to help plan the adoption of the practice. Using resource personnel here is likely to be a mistake. Their use is implied by their title—a resource, an aid in teaching, and not a substitute to take over. Sometimes a busy teacher over-uses outside people and the class loses some of its cohesiveness, its morale, and its willingness to be constructive. It begins to show the lack of a leader and fails to function effectively as a group. The essential cooperative relationship between teacher and students and among students disintegrates. A few poorly chosen resource persons, improperly used, can hasten the process. The implications are to space resource personnel properly throughout the course, and to secure resource personnel who can make contributions. This means that promotional agents, propagandists, and the like are not indiscriminately used. Teachers must be zealous of their responsibilities to the students in their classes.

The wide-awake teacher can avoid the over-use and misuse of resource personnel. When he reviews the program of instruction, months ahead of the classes, and determines the sessions where such resource assistance can be advantageously included, he should prepare a list of people who can adequately fulfill the needs designated in the program of instruction. He may find it helpful to consult with other professional people to assist him with the selection. When he is satisfied that he has identified the best that are available and that they will be able to meet the particular requirements, he should begin immediately to make the necessary arrangements. This should be done three, four, or more months ahead of the time they are to appear on the program. Good resource personnel are busy and have heavy schedules. When contacted early, they are more likely to accept the request. Whether the contact is made by phone or letter, an acceptance letter should follow, restating the essential details of the agreement. Then several weeks before the date of the meeting, another call or letter will make certain the appointment was not overlooked. This rechecking of the details agreed upon previously gives assurance that there are no misunderstandings about the agreement and that the guest teacher will be at the meeting.

In making the arrangements, the teacher's knowledge of the person to be used will determine the role he is to have in the meeting. Certainly the resource person should know the time and exact place of the meeting, the number of students in the class, the precise job of teaching he is to do, the program for the evening, the kinds of questions he will likely be asked, and the closing time of the meeting. The teacher should know the exact costs that are involved including the fee, and whether provisions are necessary for travel, meals, and lodging. He should also know what materials and aids are to be used and have them ready for the meeting.

GETTING THE CLASS READY

The class should be alerted from time to time about what is ahead in the program of instruction. In this way they should know when resource personnel will appear. Members frequently like to dress a little better when they have guest teachers. In addition, this is often the special occasion for them to invite their neighbors and friends. This, of course, is to be encouraged because at these meetings as well as for all meetings, there should be a good attendance.

The teacher should prepare the class to properly host the guest speaker and then assist so that this responsibility is carried out. The guest should be met at the arranged time and place and helped with his equipment. After being introduced to the members of the class who are present when he arrives, he should be presented properly to the entire class. When the meeting is started the guest teacher should then be given the opportunity to do what he was asked to do. The members should treat the guest with respect and extend at all times the courtesy that good manners will dictate. At the close of the meeting, he should be thanked for his contribution, paid if this was in the agreement, and helped so that he may go on his way.

The teacher has another responsibility in getting ready, in that he plans the class session so the resource person can be used effectively. This includes preparing the class members so they assume their proper roles in the teaching-learning situation. This varies, however, with the kind of resource personnel employed. To advantageously discuss these roles, they are presented in relation to each of the more common kinds of resource personnel used with adult classes.

SPECIALISTS AND EXPERIENCED TEACHERS

Such individuals are former and present teachers of vocational agriculture, teachers of other high school subjects, and extension personnel. Generally they know their subject well, they have taught adult classes, they are familiar with the adult programs in agriculture, and they are competent teachers.

These people can teach a class. If they are given an assignment their judgment on the procedure to use should be respected. They in turn will suggest how the class should be prepared for the meeting. The teacher should know this and act accordingly when planning with the guest teacher the details of the assignment. His responsibility at the session is to keep the discussion on the subject, and to either stimulate student participation or to discourage over-participation. The teacher and the class should be conscious of the shortness of a session and the enormous amount of information that these specialists can present. Consequently, every moment should be used to advantage. Here is where a tape recorder can be used. Then at a later meeting, the presentation can be reviewed and discussed.

TECHNICAL SPECIALISTS

Engineers, veterinarians, bankers, physicians and lawyers are among those in this category. They can assist in the teaching when properly directed. These people are usually not accustomed to teaching and it frequently is difficult for them to appreciate and understand the objectives of the lesson. Their thinking is often above and too advanced to relate easily with members of the class. A good way to use these people is to prepare a list of questions for them to answer during the class session. Give a copy of these questions to the guest speaker in advance of the session. At the time of the meeting when he is introduced to the class, the teacher should promote a logical order of the questioning from the class. This will keep the discussion in bounds and not digress to an exposition unrelated to the objectives of the lesson. The ability of the teacher to control the situation, to keep the discussion on the level of the students, and to utilize the resources of specialists will of course determine the results secured.

SKILLED LAY PEOPLE

Farmers, mechanics, servicemen, and businessmen are included in this group. For these people the role of a teacher could be frustrating. Their work is dealing with things and individuals rather than with groups of people. They could easily turn down an invitation to appear before the class. The teacher needs to allay their fears by telling them how they are to help and that they have a worthwhile contribution to make.

Here, as in the discussion on the uses of specialists, the teacher with the class should prepare a list of questions to be asked. A copy should be provided to the participant. Often these men are best used on a panel with the teacher as a moderator. In such a setting, they are more at ease and react more freely. Generally they talk on the level of the adult students, are aware of the common problems, relate well and can be very effective when properly utilized by the teacher.

Another use of skilled lay people is in giving demonstrations. In this they are experts technically but often not professional teachers when left on their own. Again, the regular teacher should direct the procedure so that the maximum returns can be secured from the assistant.

The contribution that these men can make in adult courses should not be overlooked. These go beyond being a resource teacher in class; they are resources to students after the sessions. They usually are from the community and are quite well known. They support the school and its adult programs. They are interested in promoting a better agriculture. They appreciate being recognized. They should not be excluded when planning adult programs of instruction.

Tests

Tests can be used with adults if properly administered. Adults having a cooperative attitude and the proper rapport with each other and their teacher will pursue efforts to learn if what is asked makes sense and is not embarrassing. The teacher can determine what aspects of a lesson students think should be emphasized by a short pre-test requiring yes and no answers. These tests with questions not requiring much writing, or questions that can be answered with a few words, are usually accepted by adults. Of course groups vary in their abilities to perform, and consequently, what may be satisfactory for one group may not be for another. Regardless, the teacher should be certain that the tests are in keeping with the abilities of the students to answer. As an example: What quantity of lime should Harry order to meet the calcium requirement for growing alfalfa? (The field is 20 acres, has clay loam soil, has a pH of 5.3 and the table of equivalents is in the bulletin, page 16.) This question is within the range of most adult groups, after the presentation, and could indicate to the teacher whether the students learned how to determine the quantities of lime needed.

Other questions could be related to the kind of calcium carrier to buy, considering high and low magnesium content and the financial advantage of limestone compared to hydrated lime. The thing to remember is that how the test is introduced and how it is used determines its acceptance and value in teaching the lesson. Formal tests and written exercises are not popular, regardless of the makeup of the group, and for older adults, who have trouble writing or who have failing eyesight, they should not be used. But, generally, tests can be a part of the teacher's usable procedures in teaching adults, when properly administered.

Result Demonstrations

Teachers, when planning programs of instruction, include lessons that teach students how to adopt tried and proven practices. However, they should also include lessons that will extend the students to consider new and sometimes inadequately tested practices. These practices, if adopted, could involve extensive changes in farm operations and the employment of machinery not now being used. The students in the teaching-learning process should be taken through these stages; awareness, interest, evaluation, and trial—but probably the final stage, adoption, may require additional information and time to adequately plan the instruction of the new practices. When the students come to the place of giving the concept a trial, it is put to the test, and can demonstrate to the students its place in their operations. Too many times, teachers fail to help their students secure the experience of seeing the new practice in demonstration, and trust that this will avail itself eventually to them by chance. With the rapid development of the agricultural industry, the competitiveness of the business, and the potential in monetary benefits accruing from the early employment of improved practices, it is not in the best interests of students to allow such things to wait. When these new practices show promise, help the class to arrive at ways to test them. Agronomic practices can generally be adopted for trial. Among those used by adult classes in demonstrating their worth are seed treatments, seed varieties, fertilizer applications, tillage operations, herbicides, pesticides, harvesting operations and storing operations. Testing practices do take ingenuity, initiative, and some time. The benefits are an earlier adoption of a trial practice and an increased economic return to the business operation. This is made easier once the student decides to adopt because he has helped in the original planning and has been following its progress during the trial. There is a feeling of extra satisfaction to both teacher and students when they can be a part of a first application of a new practice and be an authority in its use. This is a great help in stimulating adult education. The fact that the class developed a demonstration of practices, and has it to observe during its trial stage, would certainly provide the best kind of favorable public relations for the program.

Teachers who have used the technique on school farms and/or on the farms of students are its staunchest supporters. Their caution is to plan well a few good result demonstrations, conduct them so they are tried properly, and evaluate thoroughly before making conclusions because such trials and demonstrations do have limitations.

Individual Instruction

DEFINITION

Individual instruction, as used here, is defined as a teacher teaching one student. Sometimes the term is used to describe self-instruction or programmed instruction. Self-instruction occurs when a student, on his own, practices to improve his skills and when he studies to acquire understanding. Programmed instruction applies to a student pursuing a programmed procedure such as that provided in correspondence courses or in teaching machines. Both self-instruction and programmed instruction are important in teaching adults. It is an ultimate in teaching when students are motivated to go beyond the instruction in class and on the farm to learning on their own. Programmed instruction, although not a new teaching-learning device, has been receiving increased attention in the past few years. It is conceivable that in the near future, more adult learning will be facilitated by this means.

USING INDIVIDUAL INSTRUCTION

The teaching of adults in the vocations of agriculture has as its major purpose to improve those abilities to earn a living. Teaching to live a better life is, of course, not to be omitted, but the economic and financial part of living is to be stressed. Fortunately for most adult students, effective teaching will result in improved incomes. This is, of course, a desirable step to take to improve living. The key factor in achieving success for both teacher and student is effective teaching, which means that the student is performing as he was taught, and that this performance gives desirable results. Effective teaching invariably involves changed student behavior and practices. Individual instruction is often necessary for students to make a change. It is obvious that individuals differ in many ways. The problem is compounded when their situations also vary. This makes the complete act of teaching and learning almost impossible for all students when confined entirely to class instruction. The individual instruction that is needed can at times be provided before and after class sessions, at chance meetings, or by telephone. At other times, it can best be provided at the home, at the farm, at the place of business or place of application. In fact, it is imperative that a teacher devote time on the students' farms to provide individual instruction and to keep the teaching relevant to students' needs. Even the most intelligent and ingenious members of the class can use the teacher's help, and can appreciate the interest of the teacher—particularly when he calls on them at their work.

INDIVIDUAL INSTRUCTION AND ON-FARM VISITATION

Since individual instruction is often conducted on the farms of adult students, it is sometimes called on-farm visitation. This is a bit confusing in that there are also other reasons for on-farm visitation. Instruction is probably the major reason, and can be associated with some of the others. Also, in any one farm visit, many of the following specific functions may be involved:

1. Getting acquainted
2. Making a survey
3. Securing enrollment
4. Arranging field trips
5. Securing teaching materials and information
6. Planning demonstrations
7. Attending meetings
8. Making social calls

ON-FARM VISITATION TO SECURE INFORMATION

It is essential for teachers of adults to become thoroughly familiar with the home and farm situations of their students. They can accomplish this best by arranging a visit, and taking a tour of the farm with the student. During the tour, the teacher should be alert and take mental notes of what is done well, what could be improved, and how the improvements could be made. He should also note the problems that the students project. The observing teacher will learn many things about the farmstead, the home, and the family. After he leaves and before he forgets, he should record his observations and impressions. Of course, before he leaves, he should secure the specific information he needs to teach the lessons in the preplanned program of instruction for the class. This means he should be able to help the students to adopt the lesson objectives.

Another of the important outcomes of farm visitation is the developing of rapport. The personalized nature of this teaching makes it important. Students must be willing to confide in their teacher and feel that he can be of help. This information is available to worthy teachers when satisfactory relations are established and maintained.

TEACHING AND ON-FARM VISITATION

The purpose of this visitation is to complement the class work by taking the teaching to the doing level and helping to evaluate the results. It must be carefully planned so that it is effectively conducted for both the teacher and the student, and it must produce results.

The teacher should know the particular instruction that the student needs, be on hand at the proper time, and do what needs to be done. This may be to help determine difficulties encountered in adopting a practice, to help make managerial decisions, or to teach manual skills. Frequently, the student wants an evaluation of his decision and some further help in planning. The organization and procedure of the teaching are quite similar to that in class situations. However, on the farm the teaching is more informal and can be subject to distractions, interruptions, and inadequate student preparation. The successful teacher develops a procedure in conducting on-farm instruction and stays with it. That part which applies to students is made known to them. He should conduct this part of his work as if he were a physician making a house call; he is not there to visit, but to perform a task and to do his work with dispatch. He is busy, his time is important, and he has other people to serve. He must remember that his performance establishes the model and that he should assist not only a few, but all students who expect and need assistance.

ARRANGING A SCHEDULE FOR ON-THE-FARM INSTRUCTION

The fact that this kind of teaching is informal and is not specifically scheduled, as are the classes in school, offers many problems—particularly to beginning teachers. They may have time set aside during the school hours, but even this does not tell them where to go, when to go and what to do. After school hours and on Saturday, other activities seem to interfere. But teachers who are usually successful in other aspects of their programs are also engaging successfully in on-farm instruction. Apparently it is the establishment of priorities and the wise use of time that determines whether there is on-farm instruction.

TIME NEEDED FOR FARM VISITS

The time to visit adults is determined by the time available to the farmers and their teachers. Farmers can plan their work schedules for interruptions of less than an hour during most of the year. Of course, the teacher should try to avoid the peak loads during the day and the seasons. But even so, there would be only a few days during the year where time could not be planned for a short visit. A teacher of vocational agriculture, working part time with adults, could probably visit on farms one or two evenings (from 3:30 to 6:00 p.m.) a week, and on Saturday or days when school was in session and at any time convenient to the farmer, during the day and evenings in the summer months. Extension personnel often set aside several afternoons from 1:00 to 6:00 p.m. for making farm visits.

The long time visit of two or more hours poses some problems in scheduling. They can be planned for some afternoons when school is in session, but they are better suited for Saturdays and during the summer months. Farmers like the visits for the teacher's help in evaluating aspects of their program and in formulating long-term goals and plans. These calls could involve an extensive review of the records, a tour of the farm and possibly a trip to examine other farm operations.

NUMBER OF ON-FARM VISITS

To tell a teacher how many visits he should make to fulfill the purpose of adult instruction is like telling someone how he should flex his muscles when throwing a ball sixty feet. The answer, obviously, is personal. If the one throwing the ball or the teacher making the visits meet their objectives, then they are doing satisfactorily, and vice versa.

Successful teachers will make from three to six visits to the farms of their adult students each year. Their time and travel reports show that for some students they made only one or two visits during the year, while for others they made twelve or more. The program of instruction, as it varies from year to year, influences the number of visits. One year it may be the beef cattle farmers or the dairy farmers who are most vitally concerned, and it is they who then have the problems of adopting practices; while the next year it could be the farmers with other interests. One thing is certain; the good teacher will not make as many visits as he would like to make nor as many as his students would like him to make. Inexperienced teachers should be careful not to show partiality, because students are aware of the visits that are made and the time devoted to their peers. The teacher who shares his efforts satisfies his students.

PROCEDURE TO FOLLOW FOR ON-FARM VISITATION

Effective on-farm visitation requires a skillful professional teacher who does his work well and with dispatch and at the same time develops rapport with his students. Too often the thoughtless teacher leaves the impression that he has little to do and that his time is at the disposal of the student. This impression does not engender respect for the teacher or for his profession. The desirable reactions are secured when the teacher pursues his mission with positiveness, confidence, understanding, empathy, tact, and good manners. The teacher should follow the course of action that he has found, through experience, to work best. The one that follows may suggest some of the steps to consider:

1. Plan with the student the exact time and place for the visit. These appointments can be made when the teacher feels that a visit is in order. Frequently, the requests of students for calls dictates the precise time of the visit. If for some legitimate reason the appointment cannot be kept, the person who is making the cancellation should let the other party know.
2. Plan with the student the particulars he is to provide for the visit. If he is to supply materials and equipment or whatever he needs, these should be made ready by the student in advance of the visit. The teacher is too busy to permit careless preparation by the student.
3. Plan in advance the procedure to follow on the farm. Prepare an outline for teaching as well as for the other functions of the visit. The details of this outline should be in keeping with the situation. A written outline may not be needed, but it can be of help to prepare it and use it. There is nothing wrong with notes on the occasion. In fact, they may help to expedite the proceedings. Sometimes a prescription should be left with the student to assure that the correct procedure is followed; an example is in the use of sodium fluoride to control worms in swine. Getting the proper amount of sodium fluoride in a worming mixture is important.

   The preplanning also helps the teacher to remember the things he is to bring and calls attention to what he is to know and do.
4. Arrive on time and be at the place agreed upon. This is to avoid a waste of the student's time. Excuses do not alleviate this waste. Similarly, the student should be made to feel his responsibility

in this matter to the teacher. Park your car convenient to you and your mission, but also be certain that it does not interfere with the operation of the home or farm.

5. Get started on the job to be done. Certainly greetings are in order and the common courtesies should be extended, but there is a mission to be performed. That is why the visit was planned, so get going.
6. Perform the responsibilities of the assignment.
   This should be in keeping with the preplanned outline. Do this methodically and with empathy as one would expect of a professional person. Enough time should have been allowed. Be careful about extending the time so as not to be tardy for the next appointment or give the impression that time is unimportant or that the teacher does not have much to do. Get prepared to leave.
7. Make a graceful and tactful exit according to schedule.
   When the teacher leaves, the student may say, my teacher helped me, I like him, and I thank the Lord that I have such a teacher.

The teacher of adults has much to learn about individual instruction. He can make a good start by becoming convinced that it is important and learning from those teachers who are competent.

USING FARM RECORDS FOR FARM BUSINESS PLANNING AND ANALYSIS

Farm records keeping has been a part of teaching adults for many years, particularly in courses dealing with farm management. However, a special kind of program has been developed called Farm Business Planning and Analysis.

The program is growing in popularity and in many states full- and part-time instructors are employed to meet the demand. This is understandable when the outcomes from such instruction are examined. For example, farm operators who have been enrolled in programs of Ohio Vocational Agricultural Farm Business Planning and Analysis for two or more years have increased their net farm income an average of $2,608. This increase was based on net returns adjusted according to annual price changes (3).

In Minnesota, some well-organized adult programs were conducted by full-time instructors over a ten-year period. The business records of 3,518 farmers who participated in these programs were studied to determine the relationship between educational inputs and economic outcomes. Among the conclusions and implications of the study are the following (2, pp. vi-vii):

1. In a benefit-cost analysis in which all direct and opportunity costs are calculated, and where all future benefits are discounted to present value, a farmer can expect to realize about four dollars of labor earnings for each dollar of investment in the educational programs described in this inquiry. This benefit-cost ratio of 4:1 does not include benefits or returns which are nonmonetary.
2. In a benefit-cost analysis in which the benefits to the community are calculated as the aggregate rise in farm labor earnings and where the costs include the aggregate costs borne by the community, the benefit-cost ratio is approximately 2:1. This is an excessively conservative estimate since it does not include as benefits the increase in business activity which derives from expanded farm sales, nor does it include a community benefit which derives from an expanding tax base. A benefit-cost ratio which includes farm sales as a measure of business activity is 9:1. Inclusion of measures of increased tax base or other less tangible monetary benefits result in an even greater benefit-cost ratio.
3. In the first three years of management instruction, there were rapid gains in farm income. Diminishing marginal returns occurred as farmers reached practical ceilings to their capacity to employ technological improvements on existing enterprise combinations. During the fourth and fifth years of instruction, farmers reorganized and reallocated their productive resources to revised enterprise combinations. From the sixth to the eighth year of instruction, farm income increased sharply and dramatically and continued to rise at a steady rate in the remaining two years reported in this study.

The labor earnings of farmers participating in the Minnesota study are shown in Figure 7–1.

Instruction in farm business planning and analysis has as its major objective the increasing of farm income; it deals directly with the management factors that affect income. Other teaching too often is concerned with the less important and the peripheral aspects of the farming enterprise. Frequently, this teaching is incomplete because it is not taken to the doing level and students do not change their practices. In class they are likely to talk about what is good and philosophize about whether a practice will or will not work. The adult programs in farm business planning and analysis have fared better. Aside from the fact that it has produced phenomenal economic returns, it also has benefited from excellent teacher leadership. Those teachers have selected quality students who are interested in the program. They would have to be genuinely concerned to keep extensive and accurate farm records and to pay the cost of record analyses and, in many classes, enrollment fees. In addition, the inclusion of the wife in the instructional program gives it strength. The class size of

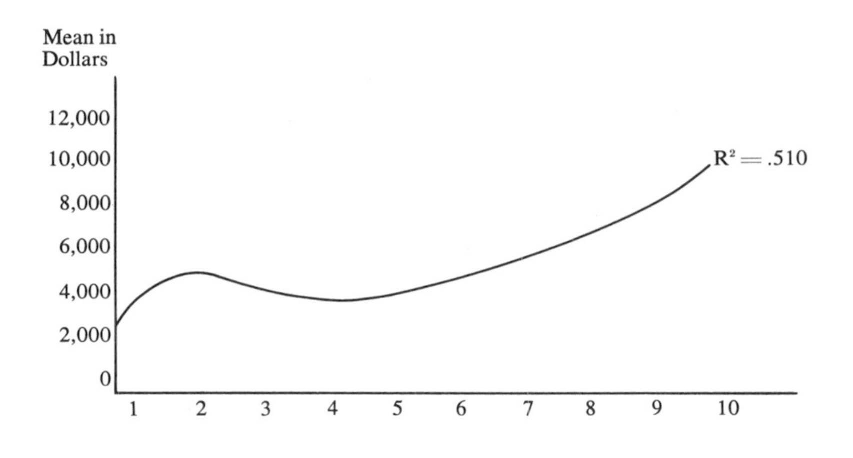

FIGURE 7—1. RELATIONSHIP BETWEEN MEAN LABOR EARNINGS AND ADULT FARM BUSINESS MANAGEMENT EDUCATION[a]

five to ten families and the nature of its instruction, much of it individual and on the farm, provides favorable influences for these adult courses.

The art of teaching adults successfully in this program is to recognize the steps to take, and then apply the skill necessary to accomplish the objectives. Programs of instruction and instructional units have been developed in many states; these should be studied and adapted to local teaching situations. Assuming that the program is approved for a course, a first step is to get farm couples to enroll. The art here is knowing the potential of the program, and then selling the idea to the kind of enrollees desired. Enrollment should be from 8 to 20, including husbands and wives. Having groups with some homogeneity in the ages of members, in the kinds of farm operations, and in social background of members will be helpful.

Keeping Records

Once the enrollment is established and the students know that keeping farm accounts is a basic part of the program, the next task is to teach them how to keep accurate records. This is a matter of developing understanding and some know-how in class, and following up with individual instruction. Here they may be taught the mechanics of record keeping and the other essentials, ranging from taking inventories to setting up a filing system.

---

[a]Based upon farmers enrolled in "well-organized" adult programs conducted by full-time instructors.

ANALYZING RECORDS

During the first year and in succeeding years, aspects of farm management are taught to prepare students to make the changes that the records reveal. This teaching is similar to the class instruction and individual instruction presented in this and the preceding chapters.

BUSINESS PLANNING

Once the records are summarized, analyzed, and interpreted, the benefits accrue only when wise planning is made effective. Thus, the success of the venture is determined by how well the planning is taught. To get this accomplished, the teacher must not only understand farm management, but be able to get students to respond by making the needed changes. These could be drastic and involve sizeable capital investments. The art of teaching to bring about such changes would certainly require a quality teacher of adults.

Evaluating the Instruction

The successful teacher is constantly evaluating his performance and looking for ways to improve the effectiveness of his teaching. After each lesson, he criticizes his performance. He may check with students to learn where and why he did well and where and why he did less than the best. He will incorporate needed changes in his next experience. As a professional person, he knows that it is up to him to seek perfection and to be worthy of his calling.

There are many considerations in evaluating. No attempt is being made here to develop sophisticated procedures to emphasize either product or process. The intent is to call attention to the importance of evaluation as a function in the improvement of instruction.

REVIEWING INTERNAL ASPECTS OF THE LESSON

The teacher of adults should look carefully at his teaching objectives and then determine why they have not been achieved. He may find the way to better results when he reacts to these questions:

1. Is the program of instruction what it should be?
2. Was the lesson really important?
3. Were the objectives properly defined?
4. Were the objectives adequate?
5. Was the subject of the lesson appropriate?
6. Was the subject understood by all students?
7. Did the students recognize their needs for the lesson?

8. Were the students interested?
9. Did the students willingly participate?
10. Did the students want to learn?
11. Did the students study outside of class?
12. Did they make wise decisions?
13. Did the students arrive at decisions appropriate to their situations?
14. Did the students show initiative in adopting decisions?
15. Did they make changes in their practices?
16. Did they make changes beyond the lesson objectives?

REVIEWING SOME OF THE EXTERNAL ASPECTS OF THE LESSON

The teacher may wish to check further to get clues on what to do to improve the instruction by answering these peripheral questions:

1. What was the attendance?
2. What effect did the lesson topic have on attendance?
3. Was the absence of certain members indicative of attitudes toward the lesson topic?
4. Did members bring others to benefit from the lesson topic?
5. Was it more or less difficult to get the session started?
6. Were the members more or less anxious than usual to continue discussion at the closing time?
7. Did the members discuss the lesson topic among themselves after the session?
8. Did the members ask questions on how to apply the lesson objectives?
9. Did the members seek teacher assistance?
10. Did the members say they were pleased with the lesson?
11. Did the members talk about the lesson following the meeting?
12. Did you really expect the students to change practices?
13. Did the members pursue participating experiences on their own?
14. Did the members seek study materials on the lesson topic?
15. What are the members saying to others about the teaching?
16. What is the reaction of the school officials regarding the program?
17. What is the public reaction to the program?

USING AN IMPROVED PRACTICE CHECKLIST

Teachers may be called upon to evaluate the performance of the class members; however, this is likely to be the exception rather than the rule. Probably for his own purpose, he may wish to know from time to time the progress made by his students. Occasionally, the teacher is asked to report the accomplishments of students. Some teachers use a checklist to secure student commitment on practices they are to adopt and to learn what the student has already adopted. The teacher should keep a copy of the checklist from each student in his notebook to assist him in planning and teaching the lesson in class, and in individual instruction. The student should also keep a copy of his checklist to remind him of his commitment and to keep him concerned with the practices he is to improve.

The checklist on page 135 has been used by teachers of adults. It is easily administered. Its simplicity merits attention.

CHECKING ON THE OUTCOMES

Teachers of adults frequently become discouraged because the effect of their labor seems so subjective. They are engrossed with day by day affairs and never take an inventory or make comparisons. Yet the work of the teacher of adults is usually much more productive than is the work of most other teachers. If the teacher has had experience in working with both high school and adult students, he will usually say that his work with adults is more rewarding than is his work with other students. This, of course, is true if his teaching is objective. The results obtained by teachers working with farmers in Farm Business Planning and Analysis Programs show phenomenally increased economic returns, and, because of records, the outcomes are highly objective. The results from other courses with adults can likewise be measured if the teacher will take the time to do it. He can get an inventory before and after on each objective he pursues in his teaching. Probably a data securing procedure could be planned to get this measurement.

There are some subjective methods of evaluation that can be significant if the teacher has been in the community for several years and is observing. He can tell whether the students are increasing their production, whether their farms are being improved, whether there is improvement in practices, whether there are more appliances in the home and whether they are living a higher standard. These and many more factors, when reviewed in relation to the instruction provided, should be really rewarding to the teacher—if he actually taught what the farmers needed to know and do.

Some Improved Practices
Emphasized in Young Farmer Program
19__-19__

Name ____

Farming Status________Acres Owned______ Acres Planted______

Used Previous to Course	Using or Will Use as Result of Course	Practices for Growing Soybeans for Grain (as defined and discussed in class and on the farm)
		1. Test soil for fertility
		2. Apply amounts of calcium according to need
		3. Buy fertilizer with an analysis based on the soil and crop need
		4. Buy fertilizer in an amount to no. acres_____ × recommended amount per acre
		5. Apply starter fertilizer in a single band 2 inches to the side and 2 inches below the seed level
		6. Select one of the following recommended varieties _________, _________, _________, _________
		7. Plant seed in 30-inch rows
		8. Plant ________ lbs. seed per acre
		9. Inoculate seed within one hour and within one-half day of planting
		10. Plant seed by ________________________
		11. Plant seed 1 to 2 inches deep
		12. Apply pre-emergence spray at the rate of __________ per acre
		13. Use a rotary hoe to break up crust and to control weed germination
		14. Use shallow cultivation
		15. Harvest soybeans when moisture content is below 13 percent—unless drying facilities are available
		16. Adjust combine according to instructions and keep losses below 5 percent

Some Improved Practices
Emphasized in Young Farmer Program
19__-19__

Name ______________________________

Farming Status________Acres Owned______Acres Rented______

No. Dairy Cattle________

Used Previous to Course	Using or Will Use as Result of Course	A. Dairy Management Practices (as defined and discussed in class and on the farm)
		1. Feed good quality hay to dairy cows
		2. Feed a balanced ration to dairy cows
		3. Purchase suitable protein supplements on the basis of cost per lb. of digestible protein
		4. Feed cows in proportion to their milk production
		5. Feed dry cows a balanced ration
		6. Supplement pasture with grain and hay during summer
		7. Grow calves according to an acceptable method
		8. Test for mastitis and sell chronic reactors
		9. Use a strip cup to detect mastitis
		10. Vaccinate calves for bangs and lepto
		11. Prevent bangs by buying from certified herds
		12. Use an approved fly-control program
		13. Sterilize milk equipment before using
		14. Cool milk to 38°F. immediately after milking
		15. Keep production records
		16. Cull unprofitable cows from the herd
		17. Use the Dairy Service Unit

Used Previous to Course	Using or Will Use as Result of Course	B. Getting Better Hay and Pasture
		1. Change cropping system to have more hay and pasture
		2. Use an approved legume-grass seed mixture
		3. Inoculate legume seed
		4. Use band-seeding method of seeding legume-grass mixtures
		5. Apply limestone according to needs
		6. Fertilize pasture according to soil needs
		7. Rotate pastures
		8. Apply manure to new seeding of grass
		9. Make hay according to dates approved for area
		C. Other Practices
		1. Keep farm accounts
		2. Adapt an updated partnership arrangement

REFERENCES

1. Bergevin, Paul, *et al., Adult Education Procedures.* Greenwich, Conn.: The Seabury Press, 1963.
2. Persons, Edgar A., *et al., Investments in Education for Farmers.* Minneapolis, Minn.: University of Minnesota Research Report, 1968.
3. Starling, John, *Mr. Farmer Do You Want To Increase Your Income?* Columbus: The Ohio State University Department of Agricultural Education, 1968.
4. Warren, Virginia Burgess (ed.), *A Treasury of Techniques for Teaching Adults.* Washington: National Association for Public School Adult Education, 1964.

ADDITIONAL REFERENCES

A Course of Study for Adult Farmer Instruction in Farm Management and Farm Business Analysis. Minneapolis: University of Minnesota, 1967.

Barker, Richard L., *An Appraisal of Instructional Units to Enhance Student Understanding of Profit-Maximizing Principles.* Columbus: The Ohio State University, Ph.D. Dissertation, 1967.

Beal, George M., *et al., Leadership and Dynamic Group Action.* Ames, Iowa: Iowa State University Press, 1962.

Boucher, Leon W., *The Development of a Farm Business Planning and Analysis Instructional Program for Ohio Young Farmers.* Columbus: The Ohio State University, Ph.D. Dissertation, 1964.

Crow, Lester D. and Alice Crow, *Readings in Human Learning.* New York: David McKay Co., Inc., 1963.

Gulley, Halbert E., *Discussion Conference and Group Process.* New York: Holt, Rinehart & Winston, Inc., 1968.

Henry, Nelson B. (ed.), *The Dynamics of Instructional Groups.* Chicago: The National Society for the Study of Education, 1960.

Jensen, Gale, *et al., Adult Education.* Chicago: Adult Education Association of U.S.A., 1964.

McCormick, Floyd G., *The Development of an Instrument for Measuring the Understanding of Profit Maximizing Principles.* Columbus: The Ohio State University, Ph.D. Dissertation, 1964.

Mathis, Gilbert L., *Managerial Perception and Success in Farming.* Columbus: The Ohio State University, Ph.D. Dissertation, 1966.

Mills, Theodore M., *Group Transformation.* Englewood Cliffs, N.J.: Prentice-Hall, Inc., 1964.

Morgan, B., *et al., Methods in Adult Education.* Danville, Ill.: The Interstate Printers & Publishers, Inc., 1963.

Rolloff, John A., *The Development of a Model Design to Assess Instruction in Farm Management in Terms of Economic Returns and the Understanding of Economic Principles.* Columbus: The Ohio State University, Ph.D. Dissertation, 1966.

Sattler, William M. and N. Edd Miller, *Discussions and Conferences.* Englewood Cliffs, N.J.: Prentice-Hall, Inc., 1968.

Chapter 8 MASS MEDIA IN ADULT EDUCATION

Contacts with mass media are so much a part of our daily living that we take them for granted. During the early hours of a new day we may watch the morning news on television as we dress, scan the newspaper as we have breakfast, listen to the automobile radio as we drive to work, and, on arriving at our desk, sort through a variety of mail which includes periodicals, circular letters and similar materials. These are examples of mass media directed toward us and important to our communication with the world around us and to the furthering of our education.

The sociologist defines mass communication as the institutions and techniques by which specialized social groups employ technological devices (press, radio, films, etc.) to disseminate symbolic contact to large heterogeneous and widely dispersed audiences (6, p. 41). This definition in its mention of "specialized social groups," "disseminating" and "widely dispersed audiences" suggests important implications for adult education in agriculture.

The mass media should not be taken for granted by the teacher of adults because they offer important advantages and the means of strengthening and supplementing his teaching. Today adult students are conditioned to the use of mass media on the part of political, social and economic groups.

Generally they welcome the use of mass media in adult education as a means of more complete and specific communication between the teacher and student.

Purposes of Mass Media in Adult Education

Mass media lend themselves to use in most aspects of the adult program of agricultural education, from initial organization of groups, to class sessions, and finally to evaluation of the program. When properly used, mass media offer important advantages to the student and the teacher. Some of these uses of mass media in adult education in agriculture are discussed in the following paragraphs.

One of the first tasks of the teacher of adult groups is to secure initial interest and enrollment. Mass media may be an effective aid in this process. Radio, television and news articles all represent means of informing potential students of opportunities in adult education. While they are valuable in themselves, they also tend to reinforce other contacts with potential students, such as personal letters, telephone calls and personal visits. Generally, the fact that a student sees an announcement of an adult class on television or reads of such a class in the newspaper lends a new importance to the activity.

By using mass media, the teacher can contact a much larger clientele in securing initial enrollment than if he were limited to personalized approaches.

Another advantage of using mass media in adult education is that it permits reaching larger audiences than can be met through the traditional classes. Mass media reach not only those who are personally engaged in educational programs, but also many of their neighbors who are provided with information as well as demonstration and example to help them in the process of adopting new agricultural practices.

Mass media are usually considered one of the most effective means of bringing about community understanding of, or developing, public relations for the adult education class. In many school districts less than 5 percent of the population of the area may be engaged in farming, and perhaps an additional 10 percent in occupations related to farming; yet we depend upon the remaining 85 percent of the population for financial and moral support for the adult education program. Mass media provide a means of keeping the public informed of our goals, our practices, our procedures and our results in adult education.

Mass media can provide valuable supplemental information for organized adult classes. Once classes are organized and meetings are under way, bulletins, pamphlets and other printed materials may be used to supplement the class discussion, providing an additional means of learning for the participants. Many of the concepts which we want to teach in adult classes concern such specifics as names of products and formulation of rations, the details of which we cannot expect students to remember. A combination of class discussion and published materials, which may be taken home for future reference, assists the student in the adoption of new farm practices.

Mass media help in recognizing individual and group achievement. There is a magic of seeing their names in print that appeals to most people. News articles, radio spots and personal mention in newsletters give an additional sense of importance to students and groups involved in the adult education program.

MASS MEDIA AND THE ADOPTION PROCESS

Mass media are recognized as contributing to the adoption of farm practices in the classical adoption model. Everett Rogers points out that information is secured by farmers at various stages of the adoption process through both personal and impersonal communication (7, p. 185). Impersonal means of communication including the mass media are most important at the early stages of the adoption of practices, according to several adoption studies. These studies point out that personal sources may be especially important at the evaluation stage in the adoption process. The implication for the teacher of adult education is to use mass media as a means of acquainting his clientele with new agricultural practices. For example, nearly all farmers in a community may be simultaneously provided with information on a new practice, such as the use of urea in feeding dairy cattle. Experience suggests, however, that mass media are more effectively used in combination with personal discussions and observations, which become the basis for the evaluation stage of the process.

The use of mass media enables a teacher or leader to maintain a continuous flow of information which reinforces his other teaching procedures during the adoption process. Studies in rural sociology show the need for continuing this flow of information over a period of years in order to secure a maximum adoption of new ideas (7, p. 110). Studies of the adoption of 2-4D as a weed spray, the adoption of antibiotics for feeding livestock, and the use of miracle fabrics in the home indicate that the early majority of adopters require from .79 of a year to 1.7 years for adoption, and that laggards require from 2.71 to 5.09 years for adoption.

These studies suggest that agricultural educators have not always allowed sufficient time for farmers to adopt new agricultural practices. Time is particularly important for farmers making those once in a lifetime decisions such as building a new pole barn or purchasing a neighboring farm. We cannot generally expect such decisions to be made as a result of one series of meetings, but during a time span of at least three to five years. During this period of adoption, mass media can be a valuable supplement, keeping students aware of new developments, of the activities of their neighbors as they try out the practice, and new information as it becomes available.

SOME ADDITIONAL PURPOSES OF MASS MEDIA

Mass media can also aid in bringing a broader spectrum of the public into the adult education program. As an example of the need for a large group being involved in adult education: let us suppose that a group of farmers were taught the importance of using a new and improved soybean variety which had produced a higher yield as well as higher oil content. In this case, the farm elevators in the area would need to stock the new soybean seed and buyers would need to appreciate the added value of the newer variety in terms of its higher oil content. If the goal of the adult education program were to secure acceptance of this practice on the part of farmers, then the success of the program would depend upon elevator operators, grain buyers, farmers, businessmen, and bankers all being made aware of this new practice and its implications for them. Mass media would represent one of the most effective means of providing this information simultaneously to the entire group.

Mass media provides another advantage to the teacher; they save him valuable personal time. While past programs of adult education for farmers have been effective for the individual farmer, they have often been far too time consuming for the teacher. Especially time consuming are such aspects of adult education as individual teaching on the farm, the teaching of small classes and the organization of small group instruction. Whenever mass media can substitute for some of these time-consuming adult education activities, they free the teacher for other aspects of the adult education program. In the past, when one extension agent or one teacher met with a group of 15 to 20 individuals and taught weekly evening classes over a period of several months by means of face to face contact, the persons reached per hour of teacher time were a discouraging few. Mass media can help the teacher or agent to teach more in a given period of time, and at the same time, permit him to teach a larger clientele.

Mass media may be used alone and unsupplemented by other teaching materials, but experience has shown that generally mass media alone are not effective in bringing about changes in behavior. An example in the field of agriculture of the use of mass media without supplemental instruction is the radio spray service for fruit growers, which the extension services in a number of states have provided. Through this service, orchardists are informed of critical times when they should spray their trees to control diseases and insects. Directions are given for mixing and applying appropriate chemicals. Many growers depend upon this information as a means of producing quality fruit, and it would seem that here is an example of the use of mass media without additional teaching. On the other hand, generally the growers who make the most intelligent and the most extensive use of the spray service are those who have gained additional insight into the reasons why it is necessary to observe certain precautions in mixing and applying a series of orchard sprays.

A KENTUCKY TEACHER USES MASS MEDIA

Charles E. Miller, who teaches adult classes in vocational agriculture at Union County High School, Morganfield, Kentucky, makes considerable use of mass media in his teaching. In a typical year he held 48 meetings with an enrollment of over 170 in his classes in farm production management, corn production, and swine and beef cattle production. Miller says, "I like to use mass media primarily because they provide an excellent source of learning and uniformity. They tend to promote a feeling of unity, effort, practice and understanding on the part of the enrollees. Then, too, they provide good resource material for public relations."

Miller continues, "I believe that a good public relations program is a major factor in the success of any educational program, especially with busy adults. I find our class members constantly discussing the program, its merits, class subject matter, and learnings obtained, and the program's overall achievements with doctors, agri-businessmen and with other business people in our community. I believe that first of all, a spark of interest in a farmer must be aroused in order for him to attend his first class meeting, then I believe I can keep him and secure results in his farming operation."

Miller uses several types of mass media, including a monthly newsletter which reaches all of his class members and is sponsored by a local bank. He uses a local radio station to announce the times of adult meetings, study hours, field trips and other events in the adult education program. He also writes an average of one article per month which is published in one of two weekly newspapers. Magazines have carried several stories relative to the program and these articles, Miller believes, help to recognize the achievement of individual members.

These advantages all point to further use of mass media in the future as a means of improving adult education in agriculture. Looking to the future, it would seem that the processes of mass media would be more available to the agricultural educator, that the audience would be more receptive because they will have a higher level of education and a general sophistication, and that, as more is known of the adoption process, there will be a greater attempt to supplement personal teaching with mass media.

New Developments in Mass Media for Agricultural Clientele

Most observers say that teachers of adult classes in agriculture could benefit by much more use of mass media in their programs. Some of the reasons for the limited use of mass media in the past have been their unavailability and their high cost. Only in the last 10 years has a variety of educational hardware become generally available in most public secondary schools. During this same period, the subject matter of agricultural education has also become more complex and has more frequently required the use of specific written information. Still another change has occurred in the clientele of adult classes in agriculture. Today's farmers and agricultural workers are better educated than they were a decade ago; they are also in the habit of reading as well as listening to oral presentations.

Commercial mass media sources in most communities include daily and weekly papers, agricultural periodicals, and radio and television broadcasting studios. The newspaper, radio or television editor generally feels at a loss in the field of agriculture because he is in unfamiliar territory. Editors are likely to turn to an "expert" in the field in whom they have confidence; this expert in many cases may be the teacher of adult classes in agricultural education.

The activities of adult education classes are of interest to the general public. Editors have a need also for performing public service for their clientele, so they are glad to carry information needed by agricultural classes. Generally speaking, more cooperation can be expected from newspaper, radio and television editors as agriculture becomes more specialized and as the general public feels a need to know more about agricultural happenings in the community.

Another reason for the greater opportunities for using mass media in adult education is that the number of agricultural publications, community newspapers, and radio and television broadcasters is increasing.

Teachers of adult classes should consider using several of these avenues of mass media in reaching their students, rather than relying upon any one means.

IMPROVED EQUIPMENT IS AVAILABLE

New resources in equipment for developing mass media on the part of the teacher represents a trend which is important to the agricultural educator. Portable tape recorders, motion picture cameras, and the office copier can be found in most modern schools. The office copier, as an example, permits the teacher in a few minutes to copy a page or two from a significant publication for an evening class. Tape recorders, which are simple to operate, permit picking up interviews with students in the class and having them played later at local radio stations.

Many schools also participate in educational television networks, which provide an opportunity for broadcasting television programs of interest to farmers throughout the state. Virginia has made profitable use of educational television as a means of teaching adult farmer classes throughout the state.

A good camera provides another means by which the teacher may make use of mass media. Pictures and slides may be used in newspapers, and slides are particularly appropriate for television programs and for presentations of adult class activities to interested community groups. Today's automatic cameras permit anyone who is willing to spend a minimum amount of time to take good quality pictures which can be used for a variety of purposes.

HIGHER EDUCATIONAL LEVELS OF FARMERS

The clientele of agricultural education has changed markedly during the past decade. An important change has been in terms of the level of education of farmers. Table 8–1 shows a comparison of the levels of education completed by two groups of farmers in the United States in 1964. These figures show that the level of education of farm operators varies from one income category to another, with the higher income groups having achieved higher levels of education. More use of mass media would seem appropriate for this higher income group of farmers.

While figures are unavailable for individual schools and classes, it is not unusual to find a sizable number of college graduates enrolled in adult agricultural education classes. During the seventies it could be expected that 50 percent of all commercial farm operators would have completed one or more years of high school, and about 10 percent would have completed one or more years of college. With this increase in students' level of education, their use of mass media, and particularly published material, can be expected to increase considerably.

TABLE 8—1
YEARS OF SCHOOL COMPLETED BY THREE GROUPS OF FARMERS

Years of School Completed	Class II Farmers (Sales of $20,000 to $39,999)	Class V Farmers (Sales of $2,500 to $4,999)
8th grade or less	29.4%	54.7%
One or more years of high school	56.0	36.8
One or more years of college	14.6	8.5
TOTALS	100.0%	100.0%

Source: 1964 Census of Agriculture

GREATER IMPORTANCE OF FARM BUSINESS DECISIONS

The audience of most adult education classes is strongly represented in a rather stratified segment of the total farming population which might be described as the upper middle class of farmers. Edward Higbee describes the upper middle class farmers as those who market commodities worth from $10,000 to $20,000 annually. Higbee says, "The 483,000 middle class farmers are 13 percent of the census total and account for 22 percent of all crops and livestock. Farm owners in this category are comparatively prosperous, for they market commodities worth from $10,000 to $20,000 annually. They are not rich but they are substantial" (3, p. 51).

Another important group of farmers served in adult education classes are young men who are getting started in farming and who are likely to be in the lower middle class of farmers in terms of income.

Planning the Use of Mass Media in Adult Education

The extent to which mass media may be used in teaching adult groups in agricultural education depends upon the clientele, the teacher's abilities and the availability of various media. The next few pages will be devoted to suggested procedures for the teacher who wishes to use mass media. Special emphasis will be given to the press, radio, television and newsletters.

In planning for the better utilization of mass media in adult education, certain assumptions can be made regarding the future of the program in which such mass media may be used. The following are some assumptions to which most teachers of adult education classes would subscribe:

1. The adoption of new and improved methods of conducting farming operations in the farm business will continue to be an important goal of the program. The key word here is adoption. Agricultural education can profit from the research of the rural sociologist regarding the adoption of farm practices. Enough is known of the time span required for adoption of practices to make some of our approaches to adult education of longer duration and also to suggest the need of mass media at appropriate times during the one- to three-year period required for the adoption of most practices.
2. A larger geographic area will be served by adult education programs. The use of area extension agents is an example of one approach to providing specialized instruction for specialized farmers. School consolidation has been an important factor in increasing the areas which are served by vocational agriculture teachers. Multiple-teacher departments of vocational agriculture, brought about by school consolidation, also provide for increasing specialization on the part of these teachers.
3. More cooperation will exist between agriculture agencies interested in agricultural education. In many cases in the past, each agricultural agency has had its own program of adult education. This diversification was probably to the benefit of the majority of the farmers up to about 1940, while they were comparatively unspecialized and as long as there were such large numbers to be reached. With the changes that have taken place in agriculture, however, the businessman-farmer has only a limited amount of time for educational purposes and he needs to spend it where he can get the most education in the most effective manner. Educational programs are likely to be more effective when they involve the efforts of several agencies rather than when each agency undertakes its own program. In utilizing mass media which may reach farmers in several counties, such inter-agency cooperation becomes particularly important.
4. The clientele of adult education will become more homogeneous. There has been an increasing tendency toward specialization in subject matter of agriculture, which has resulted in farmer specialists in different areas wanting to meet together to learn more about their businesses and their production practices. There will probably be more separate and specialized groups of adult farmers than of young farmers because for this

group there exists a core of problems common to the establishment of a new business. Adult education will also include many persons engaged in agribusiness organizations who work directly with farmers.

The case has been very well made for separate classes for young farmers because of the unique problems which they encounter in organizing and establishing a business as compared to their older counterparts.

5. Teachers of adults will become more specialized. This seems inevitable for as students of adult education become more specialized, they will demand teachers who are knowledgeable in their own areas of interest. Particularly as special full time adult education instructors are added to vocational agriculture departments, they will tend to be selected in terms of their specialized competencies in agriculture.
6. More financial support for the adult education program can be expected. The entire vocational education program has profited from additional funds which resulted from the Vocational Education Acts of 1963 and 1968. The adult education program in agriculture has benefited in some states through the use of funds which provided special teachers, teaching materials and teaching equipment.
7. It can be expected that the clientele of adult education in the future will be better educated. A very high proportion can be expected to have been in the F.F.A. and 4-H clubs and to have had vocational agriculture in high school. More and more of them will have attended agricultural colleges. Thus, their level of sophistication in agriculture will be much greater than that of previous students of adult education in agriculture.

USING NEWSPAPERS AND PERIODICALS

Optimum results in using mass media are obtained by those who are familiar with the organization and method of presentation of each of the media. Newspapers for example, have certain requirements which must be understood by the teacher of adult education. Likewise, television, demonstrations and newsletters each have their own specific requirements.

Typical of the concerns of teachers and extension agents were those expressed in an Ohio workshop on communication in agricultural education. They asked (9, p. 41):

1. How should relationships be established and maintained with newspaper editors?
2. In what form should materials be prepared for newspapers?
3. What kind of pictures should be provided to illustrate news articles?
4. How should the article be organized?

Reporting on their experience in using newspapers in their work they stated:

> The newspaper is still one of the most desirable forms of mass media. The pipe and slippers may not get used much any more, but the newspaper still holds its cherished spot at the breakfast table and beside the easy chair.
> We in agriculture need the understanding of other segments of our population. Therefore, their political support is needed for programs safeguarding the interests of all people when food or conservation of agricultural resources are concerned. If others are to support agriculture, we must keep them informed about our program. They must have a majority of our population in agreement that food is our most important product.

The following suggestions have proven useful for establishing and maintaining relationships with the local newspaper editor:

1. Through personal calls to your editor's office, find out what news of agriculture he wants and what particular news you can provide.
2. Keep him continually informed about your adult program and the current activities through visits, telephone calls, by letter, by sending him copies of correspondence, and any other appropriate means.
3. Give your editor at least two weeks' advance notice of all newsworthy events in connection with your program. This helps him to plan for adequate future coverage.
4. When you have an important event, offer to write a series of articles rather than a single presentation. Here is an example of a series of articles regarding a single activity which were much more valuable than a single story written after the field day was over:
   "Field Day for Young Farmers Is Planned"
   "Young Farmer Field Day To Be Held Next Week"
   "Young Farmer Field Day Attracts 150 Area Farmers"
5. Offer written copy either in story form or, if preferred, provide only the basic facts to a reporter, depending upon the editor's preference.
6. Offer to help your editor in arranging times and places for pictures and interviews with appropriate persons in your program.
7. Invite newspaper people to banquets and similar awards functions and give them recognition for their assistance.
8. Be of service to the newspapers even when your own personal interests are not involved. This means, for example, that there may be times when the editor needs information about a new pest in the county or unusual weather conditions, and your assistance may be of real value to him.

WRITING USABLE NEWSPAPER COPY

Hundreds of news articles pass over the editor's desk each day. The mimeograph machines of thousands of business firms and organizations turn out a grist of articles usually prepared by professional writers and illustrated by professional photographers. Your article has the advantage of being of local interest and of having been prepared by a person well known to the editor. Each article is in competition with all of the others on the editor's desk on a particular day. If your copy is to be used, however, it must meet certain requirements of the editor.

As your editor glances at a page of copy, he must be able to tell at a glance who wrote it, when the article may be used, what the article is about and how long the article should be. In addition, he needs space for writing headlines and interoffice notes regarding the disposition of the article. The form on p. 151 meets these requirements.

Writing Acceptable Copy. In writing the news story, you should follow simple accepted rules so that your copy may be in its most useful form when it reaches your editor, keeping in mind that there is no demand for news which is twenty-four hours old and remembering that your newspaper's reputation depends upon honest factual presentation of current events. Beyond this, there is no one way to write your story. Ten experienced reporters might each come out with ten different stories of the same event and all ten might well be useful and interesting. The important thing then is that you follow the rules of the game and then just sit down and start writing, and enjoy doing it.

Listed below are some general rules which should help you turn out useful and interesting copy for your news stories:

Accuracy is essential and is achieved only by gathering accurate information and double checking your final copy.

Unless you are writing an editorial, keep your own personal opinion in the background, concentrating on *what, when, where, who,* and *why*. Remember that what happened last month has much more news value than what happened last year.

Localize your story. References to local persons, organizations, and events hold your reader's interest and serve as a basis for introducing more complex and remote ideas.

Keep your articles brief and to the point. Most articles should range from two to six double-spaced pages. A three-page magazine article with one illustrative picture, a picture of the writer and headlines will occupy approximately one 8½" x 11" page.

Include tables, graphs, and line drawings if appropriate.

Leave the upper one-fourth of the first page blank, starting with your headline about 3 inches from the top of the page. In the upper right

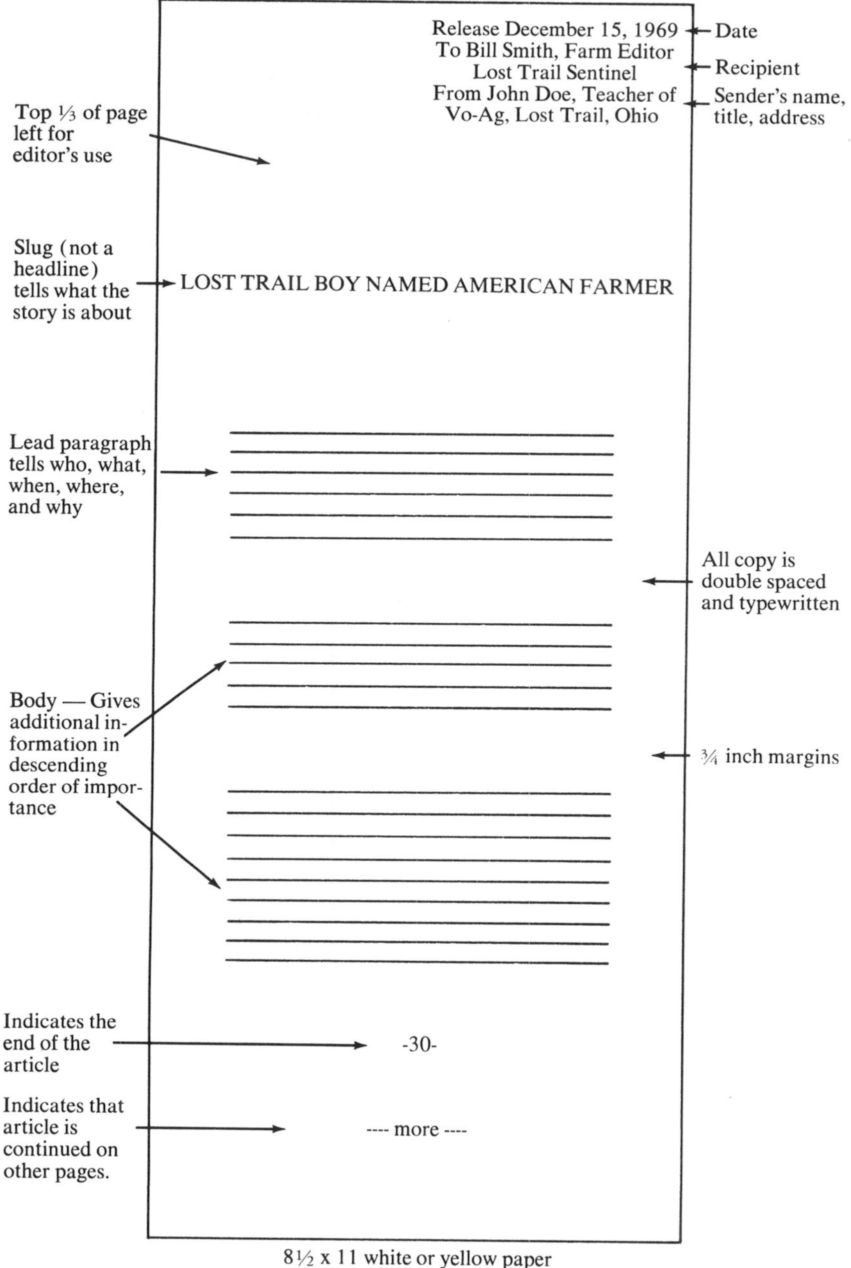

FIGURE 8—1
A DESIRABLE FORMAT FOR A NEWSPAPER ARTICLE

hand corner of the first page, give your name, the date the article was submitted, and the month in which you wish the article to appear.

All articles should be typewritten and double spaced. This also applies to titles, footnotes, and photo captions.

Writing the Lead Paragraph. The lead paragraph on most news stories should sum up the most important facts in the whole story. The lead should answer the questions who, what, when, where, and why. Further explanation and additional facts are then presented in the body of the story. Each paragraph is ordinarily more important to the story than the one below it. This arrangement makes it possible for the editor to cut the story to meet space requirements without having to rewrite it.

The humorous example below shows the relationship of the lead paragraph to the remainder of the story.

WHO ——*Willy Work,* of the Hayworth Young Farmer Asso-
WHAT ——ciation, of Hardly Normal School, will receive a *State*
WHY ——*Award for the most fantastic farm record books in the state.*

WHEN —The award will be presented *at the time of the*
WHERE —*State Round-up, which will be held at the State University, December 4, 5, and 6.*

Willy is the son of Lotta and Noe Work, of Lost Trail.

WHY —Willy receives this award as the result of four years of inaccurate work in vocational agriculture. The results of his farming program shown in record books are almost unbelievable. Willy says, "I don't believe my books would be what they are today if it hadn't been for Dad's helping me."

Nearly a million enthusiastic Young Farmers from all over the state were expected to attend this year's Round-Up according to U. B. Nice, Executive Secretary of the State Association. Nice says that current interest in farm records may have been stimulated by approaching income tax deadlines.

Making a Column Speak for You. As an agricultural leader, you can use a newspaper column to visit with hundreds of folks in your area whom you cannot see personally. In so doing, you can pass along many interesting facts and helpful ideas which will aid your overall program.

A well planned column can do many things for you which a formal news story cannot do. It can give you an opportunity to get into print bits of valuable information which would not rate a news story; it can enable you to express your own opinions on any number of subjects, and it can build for you a reputation as an agricultural authority in your area.

The following are some suggestions on writing an agricultural column from Austin E. Showman, Extension Editor at The Ohio State University (8, p. 21):

> Column writing differs from straight news writing. The most successful columns are written in a friendly, informal, chatty style. Generally, they are written in first person.
>
> You are the boss when it comes to what goes into your column. What are the main problems facing your students at the moment? What do they want to know when they drop into your office or visit with you over the phone? What interesting things did you see or hear about as you traveled over the country? At the end of a day, make note of these things and stick them in your desk drawer. They will come in handy when you start to write your column.
>
> You may use your column to announce meetings, tours, or field trips; to offer some advice or suggestions on a wide range of current problems, or to report on farm visits.
>
> Although you write your column in a conversational vein, don't get too verbose. Write in short, simple, concise statements which provide impact. Whether you are writing a column or a news story, the first sentence or two is the bait with which you catch your reader. Once you have him hooked, you can fill in with the lesser details.
>
> Most readers prefer columns which contain several short items on a variety of subjects, rather than those which comment on only one subject. There are exceptions, however, and you must be the judge on how much space you think an item is worth.
>
> You may want to talk with your local editor before you start your column. He will tell you how long he wants it and can give you some pointers on how to prepare copy to meet his newspaper's style. He also may offer some ideas on a heading or cuthead at the top of your column. A standard cuthead will help readers identify the column at a glance.
>
> Once you start your column the editor will depend on you to see that it gets to him regularly and promptly. Don't let him down.

Photographs for Use in Mass Media

Good photographs are useful in newspapers, periodicals, programs, newsletters, and on television programs. They help make mass media more interesting and lend additional understanding to the printed word. Pictures are important because they get the reader's or viewer's attention and direct it toward an idea. Most teachers are not fortunate enough to have a professional photographer available for their use, so they must be their own photographer.

The first step is, of course, to secure a camera adequate to perform the task and then to learn the mechanics of using it. Far more important than selecting equipment and learning the mechanics of taking good pictures is the task of planning and composing useful pictures.

Here are some suggestions on picture composition which may help you in securing appropriate pictures of your adult class activities which may be used in various mass media. This includes not only black and white pictures but color slides as well.

1. The picture should illustrate a single idea so well that the reader gets the point with a minimum of information in the cutline.
2. Teachers, students and others should be shown in action rather than in a posed situation. An appropriate center of interest such as a basket of corn, a weanling pig, or a basket of eggs usually adds to the picture. Pictures showing people shaking hands or pointing to objects are so commonplace that they should not be used.
3. Outside pictures are preferable to inside pictures. Most editors get far too many pictures of people in hotel rooms attending conventions. On the other hand, they get very few pictures of farm people in the great out-of-doors. When pictures are taken outside, however, the flash and sunlight technique should be used to eliminate facial shadows.
4. A limited number of people should be included in the picture. For most pictures, not over three persons should be shown, since it is important that facial expressions can be seen. This means that close-ups taken from four to six feet usually result in the best pictures.
5. The background of a picture should show some agricultural characteristics and yet not be cluttered. A barn or a silo in the background are all that are needed behind a group of people to suggest a farm setting.
6. Pictures should be appropriate to the time of year when they are used. As you collect pictures it is desirable to get snow scenes for winter pictures, pictures that show spring plowing, as well as harvest scenes in the autumn.
7. Usually pictures should be taken with light rather than dark background. This is particularly important where pictures are taken inside buildings and classrooms.

8. Pictures should usually be submitted as 8 x 10 single-weight glossy enlargements unless your editor has other preference.
9. When pictures are printed, extra contrast should be specified.
10. A brief typewritten caption describing the picture and giving pertinent information should be attached to the back of the picture with a three-quarter-inch square of scotch tape.

Publishing Your Own Materials

There are times when agricultural leaders find it most advantageous to do their own publishing. These publications may range from simple mimeographed newsletters and banquet programs to more elaborate printed brochures or bulletins containing a variety of illustrations and using color in both inks and papers.

The advantage of your own publication lies in that you become the editor. You can determine the length of your publication, including as much detail as you desire. Unlike in a newspaper story you may use color, diagrams, and illustrations. There are no limits to the quality of paper and methods of binding available to you.

Before you decide to publish your own materials, you should also consider some of the disadvantages. Most printing processes are expensive, especially for small runs. Costs per copy are usually reasonable if over 1,000 copies may be printed. Five hundred copies is often a minimum for offset or letterpress printing.

Still another factor is the professional time required. Generally, more time is required for preparation of copy than you would expect. You will need to figure, too, on the time necessary for your printer to fit your job into his schedule.

The following suggestions together with consultation with your printer should help you to publish those materials important to your program.

When the word "illustration" is used, it can mean a photograph, a painting, or a drawing. It's difficult to compare costs because quality photography and artwork are paid for on a piece-work basis. Photographs, though, usually cost less to buy than drawings.

Standardized illustration cuts may be used in your printing job. You can subscribe to a monthly service which provides illustrations for your particular printing needs. Distributors of mimeograph and multilith supplies have tracing patterns for the users of their equipment.

Line drawings emphasize simplicity and sharp, clear handling of detail. They are used extensively in multilithing and mimeographing. They summarize and suggest; the reader fills in the details. In contrast, photographs usually allow little room for the imagination.

Simple line drawings are the easiest to print on inexpensive paper and fast running letter-presses. They can be imitated readily in mimeograph and multilith work; they can be pasted easily into the copy for offset printing.

Many drawings are more interesting when shaded to give highlights, shadows, or special effects. Such shading may be put in the original drawing by the artist or added later by the engraver, following the artist's instructions.

The four main factors that govern the choice of artwork are: (1) How well will the kind of artwork complement the copy? Art and text must harmonize, and look good together; (2) Will the artwork meet the mechanical requirements of reproduction? Photographs cannot be used with the mimeograph process; in letterpress printing, line drawings look better than photos on *rough* paper; (3) Will the artwork selected allow sufficient time to carry out the ideas? (4) Can the artwork be done within the budget?

These questions help determine whether photos or drawings are best. They also guide the selection of the kinds of drawings.

PRINTING PROCESSES

Letterpress printing is obtained by pressing paper upon a raised, inked surface. The raised surfaces may be type, a cut (illustration), or a combination, and they are made of metal, wood, or rubber.

The text of your booklet (report or handbook) may be set in type by hand, or on a linotype or monotype machine. Drawings are made into line cuts; photographs are turned into halftone cuts. Type (and cuts) are checked by galley proofs; later they are checked by page proofs. The verified pages are locked into metal frames, placed on the press, tested by press proofs, and then the presses roll.

Letterpress printing can produce almost every kind of item from a postage stamp to a book.

Take a date stamp, press the rubber numbers on a sheet of writing paper, and then put your hand quickly on the impression. The date will "offset" on your hand, backwards, of course, but that is the general idea of offset printing. It is printing from a smooth (not raised) metal surface which first is pressed against a rubber blanket and then *offset* onto paper.

In simple offset printing jobs, each page is first pasted up as a unit, including such artwork as photographs and drawings. The copy may be composed on the typewriter or linotype. Proofreading should be done before the camera copy is completed, although some offset printing operations permit the use of proofs before the final copy is made.

Offset work takes less time to produce, gives softer lines, and lends itself more readily to unique layouts.

In recent years, manufacturers have developed a simplified machine for offset work called a multilith. Some of these machines are small enough to be operated in school offices. The process uses a mat (called a duplimat) which may be typed upon with a special ribbon or drawn upon by hand with a special pencil.

The multilith process is especially useful for form letters, handbills, pamphlets, folders, and small posters. Colored inks can be used by running the paper through the machine for each color. The machine is available in various sizes from the office type (handling paper 9″ x 14″) up to those meeting most commercial demands.

The ordinary mat will run 500 to 1000 copies and the so-called "long-run" mat is good for 2000 to 3000 copies. These are approximations, since the quality of the typing and the skill of the operator determine how well the mat holds up.

Mimeographing is done by a stencil sheet laid over an ink pad on a cylinder. As the cylinder revolves pressure squeezes the ink through the stencil onto the paper. The stencil usually has been cut on a typewriter (with the ribbon removed), but it may include line drawings (your own or those of the manufacturer) which are glued into a space cut in the stencil. The mimeograph does not reproduce photographs.

The mimeograph can produce form letters, small handbills, posters, programs, folders, and other small items (up to legal size typewriter paper). By changing the ink pad, various colors are possible. Up to 3000 copies can be run from a stencil, although under ordinary conditions of typing and machine operation 1000 copies is a practical working limit.

Other processes include such things as gelatin pad process, multigraph, and liquid process duplications.

Radio and Television

Commercial radio and television will be discussed together, since both of these media have a number of similar advantages: They reach a large number of people quickly and thus they are especially valuable in providing announcements of meetings and similar timely items. They give status to the program. People who hear adult education programs mentioned on radio or television tend to assign a higher prestige value to them. Appearing on such programs is a useful educational experience for members of the adult education class. Radio and television involve contacts with the entire farm family rather than just the man of the house. The number of persons reached is much greater than can be reached through personal contact. The geographic area is often increased considerably over the area served through personal contact.

The following are suggestions for commercial television broadcasts, many of which also apply to radio: Live objects should be used whenever possible in order to secure the most realism. The best live object in the adult education program is the student himself. Live and actual objects should be supplemented with still pictures, including 8 x 10 black and white pictures and color slides; 16mm motion picture film taken on farms is also useful. Short films quickly take the audience from the studio to the farm. Such films provide for transition and setting the scenes and for opening and closing the programs.

A chalkboard or a flannelgraph can be very useful for developing the program and holding the interest of the viewers throughout. Artwork should be easily understood and most drawings should be grasped within ten seconds. Lettering should be simple and large enough to read easily. A dozen words are about as many as should be placed on the screen in a single picture. Printed materials, such as extension and experiment station folders and brochures, are also good program material.

In planning your program, it is desirable to have an interest-getter during the first 60 seconds. This means introducing some unusual action or object. It might be as simple an operation as breaking an egg to show its desirable qualities, or merely a withered weed showing the result of treatment with herbicide.

While the program is being presented, the participants should look at the camera and should think of the camera as another person. They should talk at about normal speed, but keep their movements deliberate and unhurried. It helps to avoid clutter if equipment and supplies are kept out of sight until they are used. There probably will be at least one accident, which should be explained without hesitation when it occurs.

Visuals should be stationed at the proper distance from the camera and held for a period of at least 30 seconds. When participants are used on the program, the number should usually be limited to two or three persons. Remember, successful television programs are based on one basic idea and amply illustrated by a wide variety of visual presentations.

EDUCATIONAL TELEVISION IN ADULT EDUCATION

In a few states, educational television is used to bring specialists before classes of adult farmers in vocational agriculture. The Extension Service is also making increasing use of educational television.

Jewell and Oliver (4, p. 37) describe such a program in Virginia where one hundred and eighty-five teachers participated in a course in farm management which was offered in the form of ten weekly lessons. The course was sponsored by the agricultural education service and the entire cost was paid by the State Department of Education. In each of the local schools, the program was conducted under the supervision of teachers of vocational agriculture. A 30-minute telecast lesson was presented at 8 p.m. weekly and was supplemented by local teachers to make it more meaningful to the members involved. Thirty minutes prior to the telecast, teachers gave an introduction to the lesson and distributed charts and other lesson materials needed. Following the telecast, an additional hour was used for reviewing the lesson and explaining practices and procedures. Each vocational agriculture teacher was provided with supplemental teaching materials for each of the lessons. After a year's experience, a majority of teachers indicated they wanted to continue the educational television series as a part of their instruction for young and adult farmers.

Boone describes an educational television program used in North Carolina for a vocational agriculture program as follows (1, p. 31): "Television was first used in adult farmer instruction in North Carolina in February, 1966. The lessons, each 30 minutes in length, were recorded on video tape. A part of the tapes were broadcast by two educational television transmitters and the balance by commercial stations. Further use of the lesson materials was obtained by Kinescoping all 20 lessons on 16mm film and sending films to areas not covered by the television transmitters."

According to Boone, teachers usually assembled the class 30 minutes prior to the telecast lesson, and this period was largely spent in reviewing previous lessons and raising questions concerning the lesson to follow. After the telecast, the teachers spent about one and one-half hours localizing the information, reviewing the materials presented on television, and assisting with individual interpretation.

The evaluations of the teachers and students have indicated a continuing interest in this method of teaching. Administrators also support this instruction. Teachers have, in fact, gone so far as to say that this method of teaching is superior to having a special instructor present at class meetings.

Circular Letters

Newsletters, circular letters or service letters can be important to the success of a program of adult education in agriculture. Through such letters, it is possible to reach a large number of students or potential students and to provide them with personalized information when it may be of maximum value to them. The first type of letter which might be considered is that which is used to secure initial enrollment. Such letters by themselves are seldom effective, but coupled with telephone calls, visits or with other personal contacts, they provide specific, factual information on the time and place of meetings, the names of officers of organizations, topics to be considered and similar information. Such letters should not only give the facts, but they should also be inviting enough and personal enough to make people want to come to the meetings.

Kelsey and Hern say that the essentials for a good circular letter are as follows (5, p. 273):

1. An appropriate caption or salutation.
2. An unusual or otherwise interesting approach.
3. A first paragraph that will gain attention.
4. Material that will:
   a) emphasize one idea or objective;
   b) hold interest in the main idea;
   c) create realization of the importance of the problem;
   d) make an effective use of appeals;
   e) establish confidence in the facts presented;
   f) arouse a desire for the remedy suggested; and,
   g) lead to a decision to act.
5. A closing statement that will encourage prompt action. Simple illustrations and cartoons can often be copied onto service letters and add to their effectiveness.

Circular letters also provide an opportunity for feedback for the teacher. Often a card may be enclosed upon which students may indicate the progress that they are making in certain farming operations or problems that they are involved in at that particular time.

The timing of circular letters is most important. Usually such letters should be sent about two weeks before the information is to be used. Thus, a circular letter on time of harvesting soybeans should reach farmers about September 15, if October 1 is a desirable time for soybean harvest.

The circular letter teaching kit of the U.S. Department of Agriculture offers the following outline for circular letters (5, p. 273).

1. Have the purpose of the letter very clear in your mind—that is, the purpose over and above the simple objectives of getting people to read the letter. Decide exactly what action you want the reader to take:

   a) reading an enclosure
   b) attending a meeting
   c) adopting a new practice
   d) performing a service to community or county, or country
   e) answering the questionnaire
   f) obtaining information from neighbors
   g) acting to prevent spread of insect pests or animal disease
2. Appeal immediately to personal interest with a snappy statement in the first paragraph, pointing out the importance of the problem to the person addressed.
3. Point out the idea in the lead paragraph with an illustration carrying out the idea.
4. State facts concerning the nature or seriousness of the problem. Tie it down as closely as possible to the immediate locale or pocketbook of the person.
5. Suggest what the person can do to help alleviate or solve the problem.
6. Remember that the first objective is to get the letter read, so be sure that the letter is neat and appealing to the eye.
7. Caution: Protect your professional reputation and that of the Extension Service by checking the letter and the stencil for grammatical errors, misspelled words and factual errors.
8. Above all, write as you would talk to the person you are addressing, personalize your letter by using expressions used in everyday encounters with them.
   a) Direct statements
   b) Simple statements
   c) Action words with few affixes
   d) Personal reference (the "You" approach)
   e) Appropriate anecdotes
   f) Common experiences

Guidelines for Using Mass Media in Adult Education

Much of this chapter has dealt with the need for additional use of mass media as a supplement to other approaches in teaching adult classes in agriculture, and with procedures and techniques necessary to their use. The teacher next needs to develop a long-term strategy for using mass media in his community to the benefit of his students. He needs to decide which to use, when to use them, and how they can best fit into his total educational program. Finally, he should develop a calendar of educational activities extending over a period of years into which he integrates the mass media which he will use. The guidelines suggested below should help in developing such plans:

1. The educational objectives of the teacher should provide a basis as to when and where mass media should be used. The objectives of the program thus become the starting place for planning for all aspects of the program, and mass media is but one of the tools used by the teacher to make the program as effective as possible.
2. The use of mass media should be based upon an analysis of the audience. This audience analysis should determine who the audience are, where they live, what they read, where they secure their information on agriculture, their ages, and their level of education.
3. The various mass media should be employed in terms of their appropriateness for particular functions at different stages of the adult education program. As an example of the selection of mass media for specific purposes, it might be determined that the news articles and radio would be the best means of acquainting a community with the possibility of organizing a new Young Farmer's Association. Later, to provide the young farmers enrolled with information on credit for buying farms, bulletins and mimeographs might be the most effective mass media to be used. Still later, in order to acquaint the public with the results of the program, the teacher might utilize a feature article in a magazine supplemented by an illustrated brochure showing the results of the program on some of the farms in the community.
4. Mass media resources available to the members of the class should be inventoried and utilized. In most communities there is a variety of mass media available for the use of agricultural educators. Newspaper editors, radio and television personnel, and others are usually happy to provide advice and consultation and in addition publications in the form of extension bulletins, experiment station bulletins or commercial bulletins. Certain equipment may be available at the school, the extension office or in other places for duplicating or reprinting materials. All of these need to be inventoried and additional equipment needs to be secured if necessary.
5. Personnel from the mass media should be involved in agricultural education programs. The personnel of commercial mass media in most communities are interested in helping with adult education. They play a big part in informing the community of what is going on in adult education and they also, through their advertisers, assist in the program through providing the products which are part of the education program. Some plan for ac-

quainting them with the adult education program is essential, for as they become involved, they often become important supporters in the program.

6. The teacher should prepare, distribute, and use appropriate materials. This guideline suggests that teachers need to become adept at preparing materials for newspapers, radio, for writing newsletters and similar materials. Good workmanship is important, as well as an understanding of such equipment as cameras, tape recorders, and similar devices necessary to the use of mass media.
7. Evaluation should lead to adjustments and improvements as the program proceeds. This means that the teacher needs to be constantly checking with his students as to the importance, use and value of the mass media which are employed. There is not much point in writing news articles which people cannot read, or producing radio programs at a time when the clientele do not listen.

REFERENCES

1. Boone, J. B., "Adult Farmer Education by Television," *The Agricultural Education Magazine,* Vol. 41, No. 2 (August, 1968).
2. *1964 Census of Agriculture,* U.S. Census of Agriculture.
3. Higbee, Edward, *Farms and Farmers in an Urban Age.* New York: The Twentieth Century Fund, 1963.
4. Jewell, Lloyd M. and J. D. Oliver, "Farm Management Production via Educational Television," *The Agricultural Education Magazine,* Vol. 41, No. 2 (August, 1968).
5. Kelsey, Lincoln and Gannon Hern, *Cooperative Extension Work.* Ithaca, N.Y.: Comstock Publishing Associates, 1955.
6. "Mass Media," *International Encyclopedia of the Social Sciences.* New York: The Macmillan Company and The Free Press, 1960.
7. Rogers, Everett M., *Diffusion of Innovations.* New York: The Free Press of Glencoe, 1962.
8. Showman, Austin E., *The Good Word.* Columbus: The Ohio State University Cooperative Extension Service, 1965.
9. Woodin, Ralph J. (ed.), *Better Communications in Agricultural Education.* Columbus: The Ohio State University Department of Agricultural Education in cooperation with The Ohio Agricultural Extension Service, October, 1959.

Chapter 9 ADULT STUDENT ORGANIZATIONS— EMPHASIS UPON YFA

In most adult education programs in agriculture, it is recognized that students learn by participating. Students should be involved in all phases of planning and conducting the program. Most successful programs make use of planning or advisory committees to meet the interests and needs of those being served. The concept of involvement implies that there is some kind of an organization of the group. Generally an adult program in agriculture is not the teacher's program, it is shared with the students.

Primary emphasis in the development of behavioral changes in most adult programs in agriculture is related to making a start and advancement or improvement in some phase of the occupation. The programs are vocationally oriented with much consideration toward the economic improvement of the situation, but they also cover many other problems of farmers and other agriculturalists. They include establishing a home and family, participating in the development of a community that is a good place in which to live, and engaging in a social and recreational program that makes life more interesting and worthwhile. Most adult teachers in agriculture realize that these concerns and problems must be met in the development of satisfactory citizens and agriculturalists. Therefore, if opportunities are not provided to meet such needs, it is reasonable to believe that a group, as it meets together over a period of time, will want to give expression to a program that is more comprehensive than discussions concerning agriculture.

Sometimes young farmers become engaged in a program of young farmer education primarily because they want to play basketball or participate in other recreational activities after the meetings. When this interest has been met, the young farmer will often broaden or change his interests to pursue the instructional objectives—particularly if the teaching is effective.

When a group has met for a period of time, someone is likely to suggest that the meetings include refreshments and/or a social and recreational activity. The wise instructor will generally honor such an interest and request, but will place responsibility with the group. He can do this by forming some kind of an organization, informal as it may be, to make decisions and take appropriate action to accomplish the objectives.

The Value of an Organization

An organization can be valuable in many ways, including its advantage as a means of learning and as a way of developing and conducting a broader program. Most people become associated with an organization because they feel that the association will be advantageous to them. The advantage might be in satisfying their own personal desires for status or prestige, or it could be in increasing their total development as persons. In some cases they are motivated by a desire to make a contribution to community development. Hall agrees (2, pp. 146-47):

> People join groups for a number of reasons.
>
> 1. First, because it is the thing to do; it is expedient. At a low interest level the reason may be merely to conform, but at a higher level it may be because the things people want are found in groups and the things they want done are done by groups. Actually our society operates on a group basis.
> 2. Second, because it provides status, superiority and prestige or a chance to extend one's ego. A joiner so oriented has little interest in the contributions he can make. His interests are mainly in the rewards he can reap.
> 3. Third, because it provides an opportunity to extend one's vision, to enrich one's circle or to serve one's fellow. These members are less concerned with rights than responsibilities.

More specifically, the values of an organization for a young farmer group are summarized as follows (3, p. 64):

> An organization of young farmers serves as an instrument for holding the group together and for coordinating the various parts of the young farmer program. It helps to maintain the member's interest, promotes group loyalty and stimulates enthusiasm in education and farming. The organization helps to keep the young farmer program alive and insures its continuity from year to year. Also it provides the opportunities needed by the members to practice and develop their leadership potentialities. Such training and experience will enable them to assume dominant roles in leadership and policy determination in established farm and community organizations when the opportunity arises.
>
> An organization makes the teacher's work with young farmers more effective and relieves him of many minor details and responsibilities. The teacher-member relationship is strengthened and the enrollment and attendance problems are minimized when such needs are made a responsibility of the group. An organization also gives the group identity, provides appropriate awards to members for outstanding achievements and serves as a source of much favorable publicity.

The values that Hunsicker emphasized apply primarily to a group of young farmers who generally have not advanced to leadership positions in adult organizations. For adult farmers, some of these values and needs are met through active participation in such organizations as the Grange, Farmers Union, Farm Bureau, and/or Farmer Cooperatives. Other adults in agriculture are associated with enterprise and business organizations such as Garden Club, Nurserymen's Associations, Grain and Feed Dealers, Machinery Dealers, Swine Producers, and Food Processors. It should be understood that an organization is valuable to the extent that it serves its members and the members participate actively in contributing to the group.

Unless the organization serves its members or provides some values to them, and the members in turn contribute to the organization, there is little need for an organization.

> A group organization is a two-way proposal and no group can long survive unless (1) it serves its members and (2) its members serve it. Individually group members assemble for a variety of reasons. Some may want security, some recognition, and others may want to do something they cannot do alone. The group must serve these needs. On the other hand the group needs must be met; it needs a plan, it needs roles, it needs facts, and it needs action. These things the members must do for the group (2, p. 25).

Kinds of Organizations

There is no one best type or kind of organization that will meet the needs of all groups. The organization that best fits any local situation must be based upon the interests and needs of the group, as well as the purposes that should be accomplished through the organization. Generally speaking, there are two types of organizations—informal and formal.

INFORMAL ORGANIZATION

As indicated earlier, most adult education groups are organized to some extent to involve the members in planning and conducting the instructional program and the related activities. A committee selected either by the teacher or by the group to perform such functions could be characterized as being informal. In most informal organizations, the officers may be limited to a chairman and perhaps a secretary. The chairman presides over the group only when some decision is to be made with reference to an activity that should be pursued. This is convenient and helpful to the adult teacher, because under such conditions he can perform his role in an advisory capacity rather than serving as a presiding chairman. Most adult education groups in agriculture should be organized at least to this extent.

In the formation of new instructional groups, the planning committee is most often appointed by the instructor or the agency that is promoting the program. After the program has been under way for some period of time, then the group logically will nominate and elect their own planning or executive committee.

> The informal organization is characterized by the simplicity of its operations, the local scope of its influence, its dependency upon the teacher, and the correlation of its activities with the instructional program. The group recognizes that the primary purpose of the organization is the maintenance of class instruction. The organization's activities are of secondary importance but definitely contribute to the total program. The teacher is recognized as being responsible for providing the instruction and the members cooperate by assisting him with it. The group's officers may open and close each class session, and the group may hold a short business session before or following the instructional period (3, p. 65).

FORMAL ORGANIZATION

As informal organizations add activities to their program, they need to develop more rules, procedures or policies to be followed. Likewise, as membership and/or activities increase, it is necessary to devote more time to organization and the functions of the group.

The formal organization is characterized by a greater number of meetings per year, a more extensive program of activities, and a more detailed outline of operating procedures. Officers and members would assume greater responsibility for conducting group meetings, and teachers would devote most of their efforts to instructing and advising.

The Young Farmers' Association, which is featured in the remaining portions of this chapter, is an example of a formal organization. The principles and methods presented are appropriate for the development of other organizations in agriculture whose purposes are primarily educational in nature.

Organizing a Local Young Farmer Association

A local young farmer association is an organization of young men enrolled in an instructional program in vocational agriculture. The YFA, as it is commonly called, enriches the program of agriculture by providing opportunities to participate in leadership, cooperation, community improvement, social and recreational activities. It does for the young farmer what the FFA does for the high school student of vocational agriculture (5, p. 3).

Membership in a YFA is usually limited to out-of-school young men who are not more than thirty-five years of age. In some cases the age range is extended, with the provision that eligibility to hold office or participate in an awards program is limited to those members within the age limit. This is done in order to keep the YFA basically a young farmers' group. There is a tendency for groups to attain an older average age of members and not invite or attract younger members each year.

The YFA, as the name implies, includes men who are primarily engaged in farming. In some cases persons who are in agricultural businesses or who have a special interest in agriculture are invited to participate. Such matters are decided by the members of the group. As indicated earlier, the organization gives the group identity, provides an opportunity to conduct activities that are impossible by individual effort, broadens the program of agricultural instruction and assists the teacher in doing more effective teaching.

PROCEDURES IN ORGANIZING

There is no one best way to become organized as a young farmer association. Generally, there is one of two situations existing in the local community. Either a group has been involved in an instructional program in agriculture, or the organizing of a YFA represents the beginning of a program for young farmers. If a group has been meeting in a program of instruction in agriculture, the YFA would probably grow out of an interest and need to expand the program. In most cases this would be the preferred method, since the basic purpose for serving young farmers or any adults is to provide an instructional program. In other words, you first develop the instructional program and then organize in order to broaden it. In support of the other situation, there are teachers who believe that you can best reach young men or other adults by offering them a broadened program from the beginning. There is no reason why such a procedure would not be workable.

Regardless of the situation existing, there are a number of essentials to be met if the organization is to be formed and serve the purposes intended.

Teachers Become Informed. The teacher of vocational agriculture is a key person in organizing the YFA. He needs to be interested and informed about the possibilities of such a program. He should take advantage of opportunities to visit other schools where such a program is functioning, secure materials from state staff, and discuss program possibilities as well as operating procedures with those who have been successful. As he attempts to serve young farmers, he needs to be sensitive to their interests and needs so that he will know that there are some good possibilities for developing such a program before proceeding with it in detail.

Secure Approval and Support from Administration. The young farmer association should be affiliated with the local school. Therefore, it should operate within school policies. The teacher should discuss the possibilities of a YFA as well as some of the operating procedures with the school administrators. He should relate the need for facilities and scheduled activities as a part of the school calendar in carrying out the program. Agreement should be reached concerning the availability and use of teacher time.

Ascertain Interest of Young Farmers. The possibilities for developing a YFA should be discussed in some detail with young farmers individually and as a group before proceeding with any organization. A fairly complete list of the young farmers in the area or district to be served should be developed. A number of teachers have reported that they need a solid nucleus of eight or ten interested men before proceeding.

Young men engaged in an instructional program could profit from visits of a state supervisor of vocational agriculture or some members of an established YFA to present program possibilities. Likewise, young farmers in the local situation could attend YFA meetings at a local, area, or state level. Materials such as a *YFA Manual* or a newsletter issued by the state association could be distributed from time to time as another means for the prospective members to learn more about the program.

Use a Key Planning Group. If the instructional group has a Planning Committee, the teacher should work with this group in identifying some of the steps necessary toward establishing the YFA. If such a group does not exist, the teacher should select four to six key young men to serve as a Planning Committee. They could arrange for such details as making a list of possible members, setting the date and planning for an organizational meeting, identifying activities, publicizing the organizational meeting, and inviting prospective members to attend. The teacher should delegate much of these responsibilities to the members, remembering the importance of student participation.

The Organizational Meeting

The organizational meeting should be planned in detail to provide information about program and operational procedures. All persons should be welcomed and introduced. Either the teacher of agriculture or the chairman of the Planning Committee should serve as the presiding temporary chairman. During the meeting such questions as the following should be discussed and answered:

> What kind of activities would be feasible for us to include in our program? (These should include those of an educational nature, as well as social and recreation, community service, leadership and cooperation.)
> What are the possibilities for programs in cooperation with other groups, including the state association?
> Who is eligible for membership?
> How might we finance the program?
> What would be a workable meeting and activity schedule?
> Who are some others who may be interested in this program?
> What constitution and by-laws, if any, should we have?
> Should we organize? When?

If, after this discussion, there is interest in organizing, then it is logical to proceed with the election of officers. There may be some concern at this point about what officers would be needed. It may be assumed that a president, vice president, secretary, treasurer, and reporter would be desirable. However, if there is some hesitancy or concern about the need and usefulness of such officers, a president and secretary could be elected, with the remaining slate elected after the constitution and by-laws are adopted. After electing the officers, the group should establish the necessary committees. Some good possibilities at this point would include committees to develop a constitution and/or by-laws and a program of activities. A schedule for additional meetings should be developed.

CONSTITUTION AND BY-LAWS OR POLICIES OF OPERATION

All organizations, particularly those of a formal nature, should have a written set of objectives and operating procedures to be followed. In many organizations these are referred to as the constitution and by-laws. A constitution is generally thought to be a system of the more fundamental principles to be followed. It outlines the structure and makeup of an organization. Even though the provisions are broad and somewhat general, they provide a description of the way a group is organized and the manner in which the program should be conducted. The constitution is designed to serve over a period of time and is not easily changed.

By-laws are operating guidelines of secondary importance and apply generally to the local situation. They include such items as duties of officers, meeting schedules, and the setting of dues.

Purposes. Most organizations are formed to serve a special purpose or purposes which should be indicated in the constitution. Such purposes become guides for the kind of program that should be developed and imply to some extent the methods to be followed. Members and others are interested in learning about the distinguishing characteristics or special contribution of the organization. The YFA, for example, does not duplicate or replace such agricultural organizations as the Grange and Farm Bureau. It should emphasize the importance of an educational program for farmers with related activities for the development of leadership, cooperation, and good citizenship.

Another major purpose of the constitution and by-laws is to serve as a guide for action. For example, who is eligible for membership? What officers are needed? How shall they be elected? Who is eligible to hold office? These are a few of the questions that need to be agreed upon in advance so that understanding will exist. We should not make such decisions on the spur of the moment based upon personalities involved. The constitution and by-laws assure an organization of some continuity and stability in the program.

If the constitution and by-laws are to serve the purposes intended, they need to be written and followed. It is good practice to provide new members a copy of the operating policies in order that they will have an understanding of the organization before they accept membership. Likewise, new officers and/or new teachers should have copies available to guide their action.

Developing the Constitution and By-Laws. The general purposes and procedures for operating an organization such as the YFA should be discussed and agreed upon by the entire membership. However, the specific statements in the constitution and by-laws should be developed by a smaller group such as an Executive Committee. This committee should review the state constitution and by-laws, as well as those that have been used by successful local chapters. Some of the provisions to be included in the by-laws are the following:

> Who is eligible for membership?
> What officers should be elected? When should they be elected?
> How shall the nominating and voting be conducted?
> What provisions should be made for dues? How are they set and when shall they be collected?
> What meeting schedule should be followed?
> What constitutes a quorum for the transaction of business?
> What is the policy of the chapter with reference to endorsing commercial products and engaging in legislative activity?
> How may the by-laws be amended?

Before proposing a tentative draft of the constitution and by-laws to the membership for discussion and adoption, it is wise to counsel with the school administration to make sure that the procedures outlined are consistent with school policy. Copies of the adopted constitution should be available for members and others interested in the organization.

Program of Activities

The program of activities is the basic means through which the YFA chapter accomplishes its objectives. Such a program should be interesting and meet the needs of the members, as well as make a desirable contribution to the local school and community. It must be kept in mind that the YFA is a voluntary organization; the members will support it to the extent that it is interesting and worthwhile to them. Therefore, much care needs to be exercised in identifying what is to be done and when it should be accomplished.

It is always a problem to know the most desirable extent or size of program. It should be large enough to involve and challenge all members but still not so large that it becomes burdensome and competes unduly with other interests and activities that should be pursued individually and by groups in the community. In the early stages of YFA development, it seems wise to conduct a fairly simple program. Generally, it is better to have good support and wide participation in a few activities than to have a more extensive program that is not well planned or conducted.

Teachers of vocational agriculture need to advise the young farmers from time to time that the primary objective of the association is that of agricultural education. Other activities should be selected that add interest and enrichment to that basic objective. In most cases young men are quite interested in social and recreational activities. These are important, and should be included in the program, but undoubtedly the purposes of the association include something with reference to leadership, cooperation, and community service. These, in most communities, are necessary for a balanced program.

METHODS AND ACTIVITIES TO USE IN DEVELOPING THE PROGRAM

There is no one best procedure to follow in identifying the program of activities that should be followed. An effort should be made to involve all members as much as possible. It is logical to have some discussion of the entire group concerning the interests and needs of the members and the community and the kinds of activities that should be included in their program. Detailed planning should be referred to a committee. This committee should consider the adoption of some of the activities being promoted by the state association and others that have been successfully completed by YFA chapters.

The written program may take the form of a list of the activities and the month in which they should be scheduled. As the YFA matures and more members are involved in leadership responsibilities, it is reasonable to expect that there will be some activity associated with each objective of the association. In most YFAs these include community service, leadership, cooperation, and public information in addition to the basic programs in agricultural education. Some of the possibilities by such areas include the following:

Education

Conduct a year-round instructional program for young farmers in cooperation with the teacher of vocational agriculture.

Have specialists or resource persons as part of the instructional program.

Have one instructional meeting every other month on a topic of interest to both the young farmers and their wives.

Make a tour of an agricultural experiment station.
Cooperate with other agricultural organizations in sponsoring special educational campaigns.
Provide assistance to members in developing an individual filing system of useful information.
Sponsor a special meeting to learn more specifically about the school program and problems.
Devote some attention to the discussion of current social and/or civic problems.
Conduct a tour of local members' farms.
Sponsor a 150-bushel corn growing contest.

Social and Recreation

Have recreation and/or refreshments after each instructional meeting.
Take a trip to see a major league ball game.
Conduct an annual banquet with wives and other guests.
Sponsor home-made ice cream supper for families.
Have an annual fish fry.
Have Christmas party with toys provided for children.
Sponsor an annual chapter dance.
Have a bowling party.
Sponsor a fall and spring square dance.
Attend the annual state YFA camp weekend party.
Have an annual picnic and tour.
Organize a basketball, volleyball, or bowling team and arrange games with neighboring YFA chapters.
Have covered dish supper for families.
Attend area and state YFA conventions.
Sponsor a tractor pulling contest.

Community Service

Conduct farm safety campaign.
Sponsor a community clean-up campaign.
Sponsor local livestock show and serve as show officials.
Collect, repair, and distribute Christmas toys to needy children.
Assist with screwworm and Brucellosis eradication programs.
Assist or sponsor community improvement projects such as park improvement, March of Dimes, and Heart Fund campaign.
Donate blood.
Provide scholarships and other aid to outstanding rural students.
Sponsor soil testing campaign.
Sponsor projects for worthy FFA and 4-H members needing financial help.
Conduct a program of weed control.

Leadership

Send delegates to the state YFA convention.
Serve as superintendents of local and county shows and fairs.
Nominate and support candidates for areas and state offices.
Participate in the state YFA chapter contest.
Sponsor career day for high school students interested in agriculture.
Devote time to study and practice of parliamentary procedure.
Nominate candidates and assist in preparing supporting data for various awards.
Have every member serve on a YFA committee.
Assist other groups in organizing YFA chapters.
Take an active part in farm organizations.
Sponsor an extemporaneous speaking contest for members.

Public Information

Submit news articles to local paper concerning YFA activities.
Present programs before community and/or service organizations.
Present radio and television programs.
Promote an annual achievement banquet to highlight the year's activities.
Recognize honorary members elected to the YFA.
Distribute copies of the local program of activities to local businessmen and school officials.
Contribute articles to the state association newsletter.

Another step that should be taken in planning the local program is to make sure that the activities pursued are consistent with the local school policies. Therefore, the president and/or the adviser should consult with the school administration. When the program has been approved, the scheduled activities should be placed on the school calendar so that the physical facilities, such as meeting rooms, gymnasium, and lunch room can be reserved for YFA use, and so that conflicts with major school and community activities can be avoided. A program cannot be effective unless it is based upon a workable schedule.

Copies of the program should be mimeographed or otherwise reproduced and made available to each member and other interested persons. Many chapters like to include the names, addresses and phone numbers of the members, the adviser, and school administrators.

Some Problems and Cautions

As a group develops a broader program, there is a likelihood of seeing the need for engaging in some community or agricultural issues and perhaps becoming political in activity and influence. Without question, farmers and others should become acquainted with issues and be able to make proper decisions in supporting the action that should be taken. Group action, in many cases, is more powerful than individual action. However, there are a number of adult agricultural organizations that engage in this activity as a basic purpose of their program. Agricultural needs can be better served if there is a concentration of such activity in a few well-established organizations, rather than if instructional group organizations engage in such participation.

Generally, public school teachers and/or personnel representing governmental agencies should not take sides on an issue, but they should provide information to their students so that each will be able to participate as an intelligent citizen. It is likely that adult organizations would object to instructional groups engaging in political activity and/or performing some of the major functions that their group can best perform. Another reason for not engaging in political activity is the likelihood of such action resulting in a divisiveness—cliques, biases and jealousies—that would destroy the major purpose of providing a good teaching-learning situation. In order to prevent the young farmer association or other instructional groups from engaging in politics, it is well to establish policies of operation early in the development of the organization.

The Texas Young Farmer Association evolved policies that they recommend to local young farmer groups concerning political activities, controversial issues and endorsement of commercial firms or products. The policies are as follows (6, p. 13):

> *Political Activities.* The State Association of Young Farmers of Texas and subordinate units are non-profit, non-political organizations. The Young Farmers Constitution prohibits partisan political activities or activities designed to influence legislation on the part of the State Association or subordinate units. However, Young Farmers are encouraged to actively participate in the programs of farm organizations and to work to secure desirable legislation as interested individuals and members of a farm organization. Such political activity should be kept outside the Young Farmers Association.
>
> *Controversial Issue.* Young Farmers need to be kept informed. If controversial issues related to agriculture need to be considered, provisions should be made for hearing both sides. The injection of controversies not connected with agriculture is to be discouraged. Introduction of controversial and political issues into chapter activities and the endorsement of any of these can only serve to alienate members and reduce the effectiveness of the chapter.
>
> *Endorsement of Commercial Firms or Products.* The State Association of Young Farmers shall not endorse any commercial product or a business firm. However, such a policy does not prohibit the State Association from making available information on a product

or on a program offered by a commercial firm for Young Farmers. Also, Young Farmers are not prohibited from endorsing and promoting education, civic, or community service programs of farm organizations, the Soil Conservation Service, and other agricultural agencies.

Financing the Organization

Practically all organizations need money to finance their program. The amount needed depends upon the extent and nature of activities and the number of persons involved. As programs are developed and activities are considered, it is necessary to give attention to financial needs and resources.

One of the first essentials in a sound program of finance is the development of a budget. A budget is a plan or schedule of adjusting expenses to an estimated or a fixed amount of income. Each local group needs to consider the wishes of its members so far as financing activities on a basis of individual assessments or contributions versus paying for the activity from funds deposited in the treasury. For example, a pot luck supper could be conducted without any cash expenditures on the part of the individual members or funds taken from the treasury. Having a banquet at some steak house funded from the treasury is quite a different situation. Such procedures not only affect the development of a budget but have much to do with the amount of money that needs to be raised. Budgets are usually developed through the work of a committee including the treasurer. Their report should be presented to the group and discussed previous to adoption.

There is no one best way to raise money; however, most organizations have dues. This provides an opportunity for all to participate equally. It is generally accepted that a person is not active and in good standing unless his dues are paid. A set amount of dues assures the organization of a definite amount of income. The amount of dues should be determined by the need for funds and the amount to be raised from other sources. Local units affiliated with the state association need to consider the fact that a certain amount of dues must be paid to the state organization. The setting and collection of dues should be developed as a policy of the chapter and included in the by-laws.

Money-making projects are a part of most organizations. In identifying activities to use in making money, it is well to consider a number of factors including the following (1):

a. Does it have educational or social value? In many cases YFAs look upon some of their money-making devices as good social activity. In the process of raising funds, they become better acquainted, have a good time, and at the same time feel good about contributing to the organization.

b. Is it approved by the school and community? Money-making activities should be consistent with the general policies of the school and community. It is well to select activities that do not conflict with those used by some other school or community organization. The activity should be accepted as being worthwhile. This varies by communities. For example, in some places a turkey raffle would be quite satisfactory and in other communities it would be objectionable. Somewhat similar is the practice of selling merchandise. Money can be made from such a process, but ill will may be created with local merchants.

c. Is the service or product giving value received? A group selling a product or rendering a service should be sure that those who purchase the product or are the recipients of the service secure a fair exchange. This is applicable whether we are operating a lunchstand, doing custom work, or sponsoring some kind of entertainment.

d. Does it provide a fair financial return? The money-making activity should be fair to those who are involved. We need to consider the costs of a money-making project in terms of time and effort. Other things being equal, we should select those activities that give us the most income for the amount of time spent. The only exception would be where the activity has social or educational value as well as value in terms of raising funds.

e. Does it provide participation for all or nearly all members? We believe that all members should make a contribution to the organization. This includes the raising of funds as well as participating otherwise. Therefore, activities should be selected that will not only permit but encourage all members to work.

SOME MONEY-MAKING ACTIVITIES

Some money-making activities used by YFA chapters are as follows:

Collect farm machinery and equipment not in use and hold an auction sale.

Lease some land and grow a crop of corn.

Co-sponsor a rodeo with the FFA.

Operate a concession stand at a livestock show.

Sponsor a donkey ball game.

Sponsor a barn dance.

Operate a booth at a July 4th celebration.

Sponsor a swap and auction night.
Sponsor a turkey shoot where turkeys and hams are given as prizes.
Sell homemade ice cream at a special community gathering.
Sponsor a weight-guessing contest on a bull at a livestock show.
Hold a chicken barbecue.
Sponsor a tractor pulling contest.

USE SOUND FINANCIAL PRACTICES

The YFA should follow sound financial practices. Generally, these include providing a receipt for all money taken in, paying bills promptly, preparing and submitting regular financial statements and having the books audited. In some cases the YFA should develop policies with reference to maintaining a working balance and the amount of money that can be expended by the treasurer without special authorization of the organization.

Maintaining an Active Membership

After a YFA is organized, effort should be continued in maintaining interest, securing new members, and involving the entire membership in planning and conducting the program. We need to combat the problem of being too satisfied with the program and the initial membership. Special practices need to be followed in attracting new members. The addition of new members from time to time has an invigorating effect upon the program.

Well-planned meetings and activities are necessary to maintain the interest and support of the members. In order to get all members involved in the organization, a pattern of working together should be established. This includes the delegation of responsibility through individual assignments and committee work. Many successful YFAs follow the practice of having the Executive Committee meet regularly to plan some of the details of the program, make key committee appointments, evaluate progress made, and identify new procedures or practices to be followed. The teacher of vocational agriculture as the adviser should meet with this committee and use it as a means for officer training and leadership development. Much can be done by the officers in setting the proper tone for the operation of the organization.

The following suggestions are taken from the manuals of operation of the California (4) and Texas (6) YFA associations. These are procedures that they have found to be workable in maintaining an active membership and good attendance.

Have a well balanced, appropriate and interesting program of activities.

Have interesting meetings which are educational and beneficial to the members.

Have well planned schedule—start on time and quit on time.

Send cards which give the appropriate program information.

Conduct meetings by accepted parliamentary procedure. Arrange for educational program on parliamentary procedure if necessary.

Involve as many members in meetings as possible. Designate different members to arrange for programs and to introduce resource persons.

Provide comfortable and orderly meeting rooms.

Have an annual meeting devoted to the interests of the graduating high school class of vocational agriculture.

Give one year's free membership, a lapel pin, and windshield emblem decal to graduating future farmers.

Carry out a strong program of assistance to the FFA chapter and 4-H clubs so that they will be aware of the activities of the young farmer chapter.

Have a membership night once a year in which each young farmer brings in a new prospective member.

Provide door prizes at meetings if permitted locally.

Refer items of business to committees for their study and recommendation to avoid loss of time so that meetings may be kept on schedule.

Keep the school administration and public well informed of activities, accomplishments, and benefits of membership.

Working with Other Groups

The YFA can improve its program and make a greater contribution toward the improvement of farming and rural living by working with other local organizations such as the FFA, post-high school agricultural groups, Farm Bureau, Grange, Farmers Union, Agricultural Extension, and Soil Conservation. Undoubtedly, there are many mutual objectives among these organizations that would be better accomplished through cooperation in community-wide programs. Through such cooperation, the YFA will gain further support and, at the same time, the membership will learn more specifically about the programs of the related organizations. The YFA should do what it can to promote membership in the adult groups that serve the cause for better agricultural education and community improvement.

Another possibility that should not be overlooked is to have some meetings and activities with the young adult home economics group that may be involved in a program of education. There should be opportunities for joint programs and social activities between such groups.

YFA State Associations

One of the more recent developments in the further extension of young farmer education is the development of state associations of YFA chapters. This is providing additional interest, strength, and prestige to the YFA. There are many activities at the area or state level that add to the effectiveness of the local program. Some of the events sponsored by a number of the state associations include an annual convention, a young farmer and wives camp, state or area tours, and area meetings in which an effort is made to acquaint chapters and prospective chapters about the possibilities in developing their local programs.

Most of the state associations are carrying on a program of contests and awards. The young farmers seem as much interested in competitive activity as high school youth. Some of the more common awards given on the state, area, and sometimes local basis include Young Farmer of the Year, YFA Chapter Program Award, Young Farmer and Wife Community Service Award, and Production Awards such as corn growing, soybean production, dairy, beef and swine.

A number of the state associations have found considerable interest in having the wives attend the state convention with some concurrent program of their own. This kind of activity has been stimulating to the young farmer program and has provided an opportunity for a young man and his wife to get away from the farm for a couple of days which has vacation and social value.

National Institutes

In addition to the development of more state associations, which exist primarily to enrich and promote local programs, there have been successful efforts to get state representatives together in the form of national institutes. Such institutes add prestige to the program and serve the purpose of exchanging ideas, stimulating interest, and promoting new programs. All such efforts are helpful in meeting the interests and needs of the young farmers so that they will be more efficient in their vocation and more competent as citizens.

Characteristics of Outstanding Chapters

When the mode of operation of chapters which have ranked high in the Outstanding Chapter Contest each year is examined, the following characteristics of these chapters have been noted (6, pp. 36-37):

> Program of activities covering a 12-month period is carefully planned by a committee of Young Farmers and submitted to chapter membership for approval.
>
> Educational activities are based on need and interest and set up on a seasonal basis.
>
> Services of resource personnel for programs are requested months in advance of meeting dates.
>
> Requests for assistance of resource personnel are sent out by Young Farmers and not by the Advisor.
>
> The President, officers, and committees accept responsibilities and provide leadership in the chapter. The Advisor remains in the background and provides assistance as needed.
>
> Responsibilities for chapter activities are divided among committees. Each committee plans and carries out assigned activities.
>
> At least one short course is sponsored during the year under the Vocational Agriculture Specialist Program.
>
> Educational programs include two or more tours of demonstration type activities.
>
> The chapter is represented at Young Farmers workshops, field day programs, area meetings, and State conventions.
>
> Meeting notices are mailed out in advance of each chapter meeting. An aggressive Membership and Attendance Committee is functioning. Each member is reminded of each meeting in person or by telephone.
>
> Young Farmers activities are publicized regularly via news media. At least six social or recreational activities are held each year.
>
> Families of Young Farmers participate in at least two chapter activities during the year.
>
> The chapter sponsors at least one major community service project during the year.
>
> Refreshments are served at each meeting.

Nonmembers, including key community leaders, are invited to selected Young Farmers chapter activities.

Awards are provided to honor members for outstanding achievements.

Awards are presented at a banquet attended by Young Farmers' families and by special friends.

The chapter participates in the awards program of the area and State.

Exchange visits are arranged with neighboring chapters.

Committees are appointed to study and make recommendations on important problems to eliminate long drawn-out business sessions.

Business sessions are short and held after conclusion of the educational program.

The chapter is well financed through one or more fund-raising activities per year.

Members share in responsibilities of various chapter activities through committee work.

Appropriate steps are taken to secure the participation of school administrators in selected Young Farmers' activities.

Some Chief Causes of Failure

The characteristics of chapters that fail are as follows:

Educational program and other activities are planned on a meeting-to-meeting basis instead of a 12-month basis. Good programs are difficult to arrange on short notice.

An alternative program is not held in reserve in case the scheduled program is cancelled.

Publicity on chapter activities is neglected.

Members are not properly informed of meeting dates, types of programs, etc.

The local Advisor takes another job and his replacement shows no interest in Young Farmers work.

Chapter activities are dominated by one or two outspoken members.

Activities are not held in which families of members may participate.

All work is left for the Advisor or the President.

Officers do not accept and carry out leadership responsibilities.

The Advisor is overloaded with other duties and cannot give adequate time to Young Farmers activities.

Chapter members do not participate in Young Farmers activities outside the local community and do not feel a part of a dynamic organization of young agriculturalists.

REFERENCES

1. Bender, Ralph E., Raymond Clark, and Robert E. Taylor, *The FFA and You: Your Guide to Learning*. Danville, Ill.: The Interstate Printers & Publishers, Inc.
2. Hall, D. M., *Dynamics of Group Action*. Danville, Ill.: The Interstate Printers & Publishers, Inc., 1960.
3. Hunsicker, H. N., *Planning and Conducting a Program of Instruction in Vocational Agriculture for Young Farmers*. Washington: U. S. Department of Health, Education, and Welfare, Office of Education, Bulletin No. 262, 1956.
4. *Official Manual,* California Young Farmers Association, 1962.
5. State Department of Education, Division of Vocational Education, *The Ohio Young Farmers Manual.*
6. Texas Education Agency, *Young Farmers Manual,* January 1967.

Chapter 10 EVALUATION

Definition

Evaluation is concerned with placing values on processes, procedures, outcomes and activities. It is defined in the Dictionary of Education (3, p. 156) as "The process of ascertaining or judging the value or amount of something by careful appraisal." Stufflebeam (8) defines evaluation as "The provision of information through formal means, such as: criteria, measurement, and statistics to supply a rational basis for making judgments which are inherent in decision situations."

These definitions suggest that the process of evaluation offers the teacher of continuing education a means of determining which procedures and outcomes are of greatest value, and then making wiser decisions regarding the future course of the program.

Evaluation and appraisal of the adult program in agriculture takes place during every class meeting, at every conference, workshop, field trip or visit to the individual participants. Students, members of their families, school administrators, other teachers, and businessmen make such informal evaluations.

Such evaluation should be welcomed by educators, but at the same time, should be guided by appropriate criteria so that the resulting evaluations are fair and constructive and result in worthwhile improvements. Unfortunately, much informal evaluation is made by people with insufficient understanding of educational aims. Organized evaluation can be encouraged by providing a "score card" based on the objectives of the program.

Principles

Until recently, the process of evaluation has not been as much used in adult education as in other aspects of education.

Verner and Booth (12) say that early disinterest in evaluation stems partly from the nature of adult education itself, and partly from the absence of any suitable framework or instrument for meaningful evaluation. They say the situation is changing rapidly with the development of the discipline of adult education and with the increase of systematic research related to problems, interests and concerns in adult education. They claim that while there has always been some kind of assessment and measurement, it has not been systematic, scientifically accurate or particularly useful to adult educators.

Humphrey (4, p. 107) suggests criteria which he believes to be important in evaluating adult education programs in agriculture. He says that while numbers enrolled are the usual criterion of success, to him the most important ones are: (1) enthusiasm of the instructor and the school administration; (2) regular attendance of class members; (3) continued attendance year after year; (4) number of former high school students who are enrolled in the program; (5) changes brought about on the farm as a result of class instruction; and (6) requests for advice and counsel received from farmers in the community by the instructor of vocational agriculture.

Verner and Booth (12) describe two major approaches to evaluating adult education: program evaluation and participant evaluation. Under program evaluation, they classify evaluation schemes that can be interpreted in administrative terms. They say that such schemes are rarely related or concerned with educational objectives specifically, and they do not measure achievement within the context of a single instructional activity.

Program evaluation, according to Verner and Booth, is made in four principal ways:

1. By measuring the program against a standard.
2. By measuring the program against a hypothetical concept of what a good program should be.

3. By measuring against what is done by similar programs in comparable communities.
4. By measuring participation.

They also have pointed out another form of participation measurement; that concerned with dropouts. This they consider of equal importance both to administrators and to instructional agents.

Knox (5, p. 111) makes these suggestions as major steps in evaluation: (1) establish clear and realistic objectives as a prelude to evaluation; (2) determine the type of information to be collected; (3) select appropriate methods of selecting information regarding achievement; (4) analyze the information; and (5) use the results to improve the education program.

One of the most practical statements of the principles of evaluation was made by Dr. S. S. Sutherland at a conference on Objectives and Evaluation in Vocational Agriculture, in 1966.

These principles were stated as follows (9, pp. 13-18):

1. Evaluation of educational programs should be made in terms of the objectives of these programs.
2. Evaluations should include assessments and appraisals of both product and process.
3. Evaluation should be a continuous process, not just a point in time judgment.
4. Evaluation should be made by teams composed of both professional and lay personnel.
5. Evaluations of publicly supported programs should include economic factors and should be concerned with input-output relationships.
6. Evaluations and appraisals should be made not only on the basis of what has been done but also what has not been done.
7. Major purposes of evaluation should be to provide quality control and a basis for intelligent change.
8. Evaluation should be concerned primarily, if not exclusively, with the key indicators of success or failure.

Each of these principles has implications for the evaluation of adult education which will be discussed in further detail.

OBJECTIVES

One of the most recent and important statements of objectives for agricultural education was published by the U. S. Office of Education in 1965 (10). This publication discusses current needs, purposes, the person who should be served, and the major program objectives for vocational and technical education in agriculture.

In discussing the clientele to be served, this report specifically mentions "employed youths and adults who need vocational and technical education in agriculture to upgrade their occupational performance," and "those youths and adults who cannot benefit from regular vocational and technical education programs in agriculture but who can benefit from special programs designed to meet their needs." Both of these groups can be served only through programs of continuing education.

The major program objectives for vocational and technical education in agriculture are set forth in the following six statements:

1. To develop agricultural competencies needed for individuals engaged in, or preparing to engage in, agricultural occupations (farming or ranching)
2. To develop agricultural competencies needed by individuals engaged in, or preparing to engage in, agricultural occupations other than production agriculture
3. To develop an understanding of and appreciation for career opportunities in agriculture, and the preparation needed to enter and progress in agricultural occupations
4. To develop the ability to secure satisfactory placement and advance in an agricultural occupation through a program of continuing education
5. To develop those abilities in human relations which are essential in agricultural occupations
6. To develop the abilities needed to exercise and follow effective leadership in fulfilling occupational, social and civic responsibilities.

Each of these major program objectives has important implications for adult classes in agriculture. Some of these implications are as follows:

> The second objective suggests that continuing education in agriculture should extend beyond farming. This further suggests that adult education classes should be organized to develop agricultural competencies needed by individuals interested in preparing for off-farm agricultural occupations. There has already been a start in this direction by some teachers.
>
> Objective three may add another dimension to adult education courses recognizing that career selection is a lifetime concern of most individuals. Many farmers are engaged in a second occupation and for a variety of reasons many persons leave farming to enter other occupations.
>
> Objective four speaks to the problem of placement. Placement assumes a more important role in the young farmer programs today as establishment becomes more and more difficult in agriculture because of the competition for land and resources and the large amounts of capital required.

Objective five which identifies the need for abilities in human relations also has implications for adult education in that the success of today's farmer depends more and more upon his relationships with employees, with customers, and with the persons from which he secures the inputs for his agricultural enterprise.

The sixth objective, which is concerned with leadership, gives a broad charge to the program of agricultural education. It suggests that if we are concerned only with the development of improved practices in agriculture that our base may be too narrow to meet the needs of our clientele.

Objectives for Local Adult Education Programs. The principle of evaluating in terms of the objectives of the programs requires that a definitive list of objectives be developed. Objectives of local programs should be in line with larger objectives, such as the preceding, which may be set up at the state and national levels, but at the same time these objectives must represent the wishes and desires of those who participate in the local adult education program. Some objectives of adult programs will relate to the processes and procedures used in conducting these programs. Examples of "process" objectives are those dealing with teaching procedures, kinds of group meetings and types of social activities. Another appropriate and even more important type of objective deals with behavior changes which take place in the students as a result of the program. Examples of such objectives are those concerned with changes in farming practice of the students enrolled.

The following objectives were developed by a vocational agriculture teacher for a farm business planning and analysis program. These objectives indicate clearly the behavior changes which the teacher envisions. They were stated as follows:

1. To develop an appreciation for the value of complete farm records.
2. To have each enrollee in the first year program complete an accurate farm account record of his own business.
3. To secure an accurate inventory on each participating farm.
4. To set up complete and accurate depreciation schedules for each participating farm.
5. To develop an understanding of a farm business summary and analysis.
6. To have each farmer in the class receive a computer summary and printout sheet.
7. To develop the ability to make sound management decisions based on past records and summaries and then apply these decisions to the operation of a farm business.

These objectives are appropriate, challenging and interesting to the student. In addition, they are stated in terms of such specific behavior changes that their accomplishment can be easily measured.

Extension Objectives. The Cooperative Extension Service has made a comprehensive statement of recommendations for the future (11), many of which relate closely to adult education in agriculture. The following statements are listed as functions the service must perform in its educational mission.

> The effectiveness of Cooperative Extension Service in achieving its mission will be to a large extent determined by how well the staff integrates the entire process of continuing education into an over-all strategy of education. The strategy must include (1) planning and preparation, (2) teaching, (3) evaluation, and (4) continuous staff recruitment, training, and development.
>
> Planning and Preparation
>
> Extension teaching follows many processes of planning and preparation that must be undertaken before effective teaching can begin.
>
> *Developing and Maintaining a Climate for Education*
>
> Developing and maintaining a favorable climate for education is an increasingly important part of the Extension function. Too often this climate has been taken for granted. This development must be pursued within the university as well as with the public. The university administration and staff must recognize continuing education as an important university function.
>
> Extension staffs must develop a favorable climate for extension education among the people. It will be necessary to aggressively seek out and consult with groups that heretofore have not been involved with Extension. Low-income groups, community leaders and public officials are examples.
>
> The strength of an informal continuing education program lies in its relevance to the needs of people. Extension must establish the same type of continuing communication with organizations and groups representing disadvantaged citizens, non-rural youth, and ethnic groups that has been developed with farm organizations, commodity associations, business organizations, homemaker councils and co-operatives.
>
> *Integrating Research and Extension*
>
> As Extension programs expand into new areas and acquire more depth in the old ones, there is continuing need for effective integration of research and extension activity within the USDA and the university.

These relationships now exist to a fairly high degree in the departments of the colleges of agriculture and home economics. Extension must build these relationships in other colleges in the university and between Federal Extension Service and the research agencies of the departments of government other than the USDA.

Assembling the Audience
The audience building function of the local Extension staff is becoming an increasingly important and specialized function. Those who most need new knowledge are least apt to seek it. As extension works with different groups in the future, the staff must acquire new skills in audience assembly, since techniques used for present audiences may be completely invalid for new groups.

PRODUCT AND PROCESS

The principle which suggests that evaluation consider both product and process has several implications. The "product" of the adult education program refers to the student and, more specifically, the changes or improvements which take place in his behavior as a result of his educational experience. This behavior change may be expressed in terms of increased crop yields, efficiency in livestock production or in similar measures. Such behavior changes may be quite objectively measured.

School executives and boards of education need to know what economic return may be expected from the adult education program, to justify expenditures of funds. An adult education program which has been in operation for a number of years has required considerable financial support and the community expects an economic return. Some of the returns that should be expected from adult and young farmer education are improvement in farm income and in the farming practices. Over a period of years the program should result in more efficient crop and livestock production, higher farm incomes, and a higher standard of living for farm families.

In addition to purely economic outcomes, some of the following may also be expected. Some of the best young people in the community should take their places in the agricultural industry. The young farmer program, particularly, has an opportunity to reverse the "brain drain" from the farm, by maintaining the interest of young men in farming. The adult education program should result in leadership and the development of leadership skills on the part of those enrolled as they take their place in the community. Individuals involved in the program should exhibit personal growth and development in terms of accepting responsibilities and developing qualities needed for effective citizenship in their communities.

Other outcomes may not be as easily quantified, but have even greater potential value. The ideals of the kind of family life which should prevail on the farm may be as important as the production of crops and livestock, and may be a legitimate goal of a program of adult education.

The process aspect of evaluation consists of the procedures, ways and means which were used by the teacher and the group in learning and in conducting classroom and laboratory discussions. It is much easier to evaluate the process of adult education than the product, and most evaluative instruments emphasize ways and means used by the teacher rather than outcomes which result in changed student behavior.

CONTINUOUS EVALUATION

The development of a continuing evaluation process is important. This principle suggests that evaluation should begin with the first meeting of the group, and should continue on an informal basis, but that it should be further supported through the introduction of such formal devices as checklists, rating scales, and performance tests throughout the program.

Probably more courses in adult education are evaluated at the last meeting, or at the close of the course, than at any other time. While this is a desirable time to make an evaluation, the above principle suggests that it is even more appropriate that this "end of the course" evaluation be supplemented by a continuing evaluation effort.

Sutherland's principle, that evaluation should be made by teams composed of both professional and lay personnel, raises some questions of its practicality in the evaluation of adult education programs. Certainly there are advantages of "outside evaluation" but much of the evaluation of adult education programs will probably continue to be made by representatives of the local community, together with the participants in the adult education program, and the teacher himself.

FORMAL VS INFORMAL

A review of current practice in adult education in agriculture shows that there has been little emphasis upon formal evaluation of adult education programs, but that the more perceptive teachers and leaders of the adult education program have made much use of informal evaluation such as the utilization of planning committees, having members assist in developing programs and in informal discussion with community leaders.

As competition for the educational dollar becomes keener and as a greater variety of adult education programs become available, it seems inevitable that more use of formal means of evaluation will become necessary and desirable.

It is interesting to note that as early as 1940, the Agricultural Education Branch of the U. S. Office of Education had developed an instrument for an organized evaluation of local programs of vocational education in agriculture (6).

This bulletin reports that programs of agricultural education designated as "Superior" included an adult evening school, that the average enrollment in these superior programs was 36.5 students per class, and that the average number of meetings was 18, which were held over an average of 7.1 months. Some of the criteria which were used in evaluating adult evening schools at that time were the use of functioning advisory councils, the relationship of the agricultural program to the community program of adult education, and support for the program on the part of the administrative head of the school. Other criteria included the use of a well organized publicity program and a plan for teacher visitation on the farms of those enrolled.

Teachers Need Evaluative Competence. The teacher or leader of the adult group becomes a key person in developing an organized evaluation program. Generally he is the person who sees the greatest need for the evaluation and who has the most understanding of the process.

The teacher needs to recognize the importance of program evaluation and its differences from student evaluation. He needs to be able to develop an overall evaluation plan which will fit into the time schedules of those involved, and which will include all appropriate personnel.

He next needs to understand the development of appropriate educational objectives and to prepare comprehensive and valid criteria growing out of these objectives. He has a responsibility to develop reliable, valid and practical instruments of evaluation. Fortunately, in many states such instruments have been developed by others and the task is largely one of selecting the one most appropriate. Finally, the local leader needs to scale and summarize the results of the evaluation, and communicate the results to school administrators, boards of education, and others who are concerned with the program, and in conclusion to make use of the results of the evaluation in terms of program improvement.

Planning Committees. Planning committees made up of members of the adult or young farmer class represent one of the best means of informal evaluation of the adult program. These committees should have a major responsibility for planning the instructional program for their classes as well as for planning for social and civic activities. They should meet prior to the beginning of a new program to formulate plans for the meetings, but more important perhaps than deciding upon topics and speakers for meetings, they should take a hard and close look at the objectives of the program and the extent to which these objectives are being accomplished.

Certain principles of leadership should be followed in helping such lay committees to work effectively. The members should be selected wisely with one of the major criteria being their interest in the adult program. They should be notified officially of their appointment, usually by letter from the teacher, and they should agree to accept this responsibility. At their first meeting, the teacher needs to explain their responsibilities and to help them organize under a chairman and a secretary and to set up a time schedule of meetings defining individual responsibilities. It is, of course, important that the teacher should shift the leadership role from himself to the chairman of the group. The teacher needs to make sure to recognize the contributions of the committee and to act upon their recommendations.

The committee should learn the process of evaluation and understand that this is one of their most important functions. Time should be provided for them to make their deliberations. It should not be expected that a planning committee can accomplish much if it meets only late in the evening after the regular class meetings.

During the whole process, the development of democratic leadership on the part of the individual members of the planning committee may be as important as the evaluative functions which are performed by the group.

Overall Advisory Committees. Some teachers who have advisory councils or committees responsible for the entire program of agricultural education may question the desirability of a separate planning committee for the adult program or for the young farmer organization. There is a place for both types of committees. Certainly the adult education program of a department is of such importance that the overall advisory committee should consider it as it evaluates various parts of the entire program. In some departments, such committees may also serve as planning committees for the adult education program. As the program enlarges however, and particularly in multiple teacher departments, each adult education class or activity should have a planning committee which is charged with planning the program and evaluating its outcomes.

John H. Hendrix, who is an adult farmer instructor at Hopkinsville, Kentucky, provides an example of the use of both county and class advisory committees in evaluating and planning adult farmer programs. Hendrix explains his procedure as follows:

> Our program was initiated on July 1, 1957. The first step in planning was to appoint a County Advisory Council. Members of this council are made up of the County School Superintendent; Board of Educa-

> tion members; agriculture teachers; school principals; Soil Conservation Officers; Extension Service Officers; and farm representatives.
>
> The duties of the County Advisory Committee are to set up goals and policies for both the young and adult farm program. In the early years of the program, the Council met biannually, in the spring and fall. After four years, the Council began meeting annually to revise the program where needed. Officers of this committee are: chairman, vice-chairman, and secretary.
>
> The second step in organization was to form a Class Advisory Committee for each class in the County. This committee consists of from eight to twelve farmers residing around the class center. These committee men are put on a three-year rotating basis of service. The purpose of the Class Advisory Committee is to elect class officers; chairman, class secretary, treasurer and reporter; to select the subjects to be taught during the class sessions; to decide on the time class will begin and to make recommendations for improving class instruction.
>
> The class Advisory Committee is solely responsible for selecting subject matter to be taught in the courses. An example of courses taught in a series is: three sessions on corn production, three sessions on fertilizers, and four sessions on beef cattle production. The Committee is responsible for the operation of the class. I work for them. One outstanding result of this Committee is that it is entirely responsible for class recruitment. No formal recruiting is necessary.
>
> These two committees, the County Advisory Committee and the Class Advisory Committee, are largely responsible for the success of our program.

Of particular interest in this example is the instructor's willingness to give important responsibilities to his planning committees. Perhaps too many adult instructors in the past have hesitated to delegate activities which others could better carry out.

Coordinating Evaluation Efforts. Evaluation of the continuing education program in vocational agriculture is usually more productive when such evaluation is a part of a school wide evaluation effort. In such an evaluation each activity of the school, including continuing education, is appraised in terms of its contribution to the total educational effort. Not only is the evaluation effort, but also the program planning which results from it, are likely to be well coordinated in a school-wide evaluation.

Agricultural extension agents in many states have made effective use of members of their clientele groups in evaluating the total extension program of their centers. County advisory groups often meet regularly during the year and spend a good portion of their efforts in evaluating the several aspects of the extension program. An interesting development in some states has been the organization and evaluation of county adult programs in agriculture, in which extension agents and vocational agriculture teachers work cooperatively, and subsequently carry out a joint evaluation of the program.

Adult education in agriculture is an important part of a local program of agricultural education which may include various other services and high school as well as adult programs. It is also an important part of the total school program and thus it might be evaluated as a part of either a vocational program or a total school program. Other levels of evaluation may occur at the district, state or national level. For most programs of adult education, evaluation should take place without waiting for other, larger scale, evaluations to develop.

Evaluation offers so many advantages and is so necessary to the ongoing adult education program that it should be incorporated as a part of the program every year rather than waiting until a more formal and larger scale evaluation may be mounted in the school community or state. When such a larger scale evaluation does take place, certainly adult education should be evaluated along with other parts of the program.

COST-EFFECTIVENESS EVALUATION

The principle that evaluation should be concerned with input-output relationships is a reflection on the current concern in government circles with cost-benefit and cost-effectiveness analysis. Cost-benefit analysis, which is based upon objective economic measures of costs and returns, has important shortcomings as a sole measure of educational effectiveness, since many of the more important results of education are expressed in other than economic terms.

One of the decisions facing a board of education may be whether to offer adult education courses. Other decisions which may follow include the number of courses and for what clientele. Cost-benefit analysis or input-output relationships help in making such decisions. In the case of an adult education program, it is not difficult to calculate the cost of the program, but the returns are much more difficult to secure.

Some recent studies have been made of the costs and benefits of Farm Business Planning programs. A study made in Minnesota of investments in education for farmers showed a cost-benefit ratio for the individual participant of four to one. In other words, for each dollar which the individual invested as recommended in his farm management program, he received a return of $4.19 for labor and management. The dollar value of farm sales was used in calculating increased business activity in the community and the cost-benefit ratio increased to about nine to one. The study concludes with this significant statement (7):

> Any investment with a cost-benefit ratio similar to that shown for education in Farm Business Management is a valuable holding for the community. As community action groups, boards of education, chambers of commerce, and others seek ways to build affluence in rural communities and protect local economies, Farm Business Management Education should be among the high priority alternatives.

EVALUATION AND DECISION MAKING

The principle which states that evaluations should appraise not only what has been done but also what has not been done has important implications for adult education. This principle applies to those whom we serve in adult education. The clientele of adult agricultural classes in most counties and communities has been largely made up of those whom we know best, the upper-middle class of farm operators. Teachers should give thought to what has *not* been done for young men who are becoming established in farming, for low income farm families, for part-time farmers, and for other similarly neglected clientele groups.

The same principle applies also to the course of study for adult classes. Our emphasis upon crop and livestock production has been heavy and we have taught much less about reorganizing the farm business to meet the current price-cost squeeze.

Sutherland points out that the major purposes of evaluation are to provide for quality control and the basis for intelligent change. Both of these factors are important in developing local programs of adult education. Quality control means that there is a uniformity of quality in what is taught and within all aspects of the program, and intelligent change grows out of quality control. When the standards of quality are not met in industry, an immediate attempt is made to make changes which will improve the product. Adult education can profit from this example.

The use of key indicators of success or failure are important in any educational evaluation and particularly in adult education. An example of the use of key indicators may be found in the development of a student achievement test. After a teacher has taught for a week, he may decide to test his students' knowledge of the area taught through the use of a quiz consisting of ten objective questions. During the week he may have taught a thousand different facts, concepts and segments of information, but he assumes that the ten questions which he asks represent a fair sample of all of these and hence he considers them as key indicators.

In selecting a key indicator, a teacher of adults who has taught a course on the selection of dairy sires might decide that one of the best indicators of success of his teaching would be the number of farmers who had studied pedigrees of various sires available for their use at an artificial insemination center.

Using Evaluation in Teaching the Adult Group. The process of evaluation can be profitably used by the teacher or leader and his students as a challenging and interesting method of teaching. The following is a process which involves the students and the teacher in evaluating the teaching process as it develops:

1. Involve the students in selecting the subject matter to be discussed. This means that not only should the planning committee recommend the general topics, but that the students should be asked for their suggestions as to what should be included in the discussions which are to follow.
2. Have the students decide upon the level of proficiency they believe they should acquire. It is desirable to have those levels of proficiency specifically described. For example, in a unit on draining, the use of the transit might range from understanding the value of using the transit in laying out draining lines, to being able to lay out a drawing line with a minimum of fall.
3. Secure the opinions of the class as to how evidence of learning should be measured. In the example above, if students are to be taught to use a transit in laying out a drainage line, the group might agree that all members of the group should be able to operate a transit.
4. Secure the suggestions of the class as to resources, procedures, methods which are most likely to bring about the type of learning which they believe to be desirable. At this point members of the class may contribute ideas which would not have occurred to either the planning committee or the teacher as the series of meetings is planned.
5. Appraise the progress of individuals as the teaching progresses. This is somewhat at variance with much of the advice we hear at adult education, for we hear that adults prefer not to be "tested." On the other hand, there are many types of written exercises which include achievement tests, performance tests, check lists, identification, and judging devices. More of these can probably be employed in adult education than have been used in the past. In the writer's observation, testing has contributed to the learning in many adult classes. Testing can serve as a reinforcement of learning and taking a test may develop a deeper understanding on the part of the student. Testing also contrib-

utes to the motivation of the student in that it constitutes a challenge to him. Simple written tests also help to establish in the student's mind that adult education is more than entertainment, and that the real goal is the development of abilities in the individual student.

6. Evaluate the series of meetings or the classroom procedure in terms of "if we had it to do over again."

The use of the above procedures can increase the involvement of members of the group in the teaching-learning process and can place the teacher in a role of guiding the learning process rather than in an authoritarian position of deciding what should be taught and how it should be taught.

Using Evaluation in On-Farm Teaching. Much of today's teaching of agriculture to adults is directed toward helping them to make wise decisions regarding their business operations. The success of those engaged in an agri-business as managers or employees depends largely upon their ability to evaluate and analyze facts as a basis for decision making. The success of today's farmers, therefore, depends upon their decision making ability. While the teaching of decision making begins in the classroom, follow-up on the farm or ranch is essential. Decision making requires abilities in evaluation and analysis of facts. Evaluation procedures can serve as a guide to the teacher as he teaches individual students to make sound decisions. Much individualized teaching is done through farm visits and here the evaluation process becomes the heart of a successful on-farm teaching visit.

Certain basic questions can be used by the teacher to guide a student's thinking in making a managerial decision during the course of a farm visit. Some appropriate questions are as follows:

1. *What is your goal?* Here, the student needs to be guided to state his farming goals in terms of specific outcomes such as yields per acre, production per animal, cost of production per unit, or family labor income.
2. *What progress has been made to date?* This question provides an opportunity for the teacher and the student to review what is being done at the present time and what has been done. In the case of a car lot of steers on feed where the student has expressed a goal of a gain of two pounds per day per steer, questions could be raised as to how long these steers have been on feed, what their rate of gain has been, and what amount of gain they should have made up to this time.
3. *What desirable outcomes have resulted?* Farming is often a discouraging business and often the operator can see only his difficulties. Teachers need to help their students see evidence of

progress which they may have overlooked. Students are anxious for commendation and the teacher should look for evidence of achievement and progress.

4. *What undesirable outcomes are in evidence?* In the case of the steers this might be in terms of unthrifty appearance, lack of uniformity, and a low rate of gain. At this phase of the discussion it is important that objective information be available. This is one of the advantages of teaching on the basis of farm records. The facts are there and provide a basis for evaluating the outcomes. Other means of taking a reading on the success of farm ventures include: taking weights of cattle, counting stands of crops, estimating crop yields, and weighing milk and other livestock products. Sometimes such evidence can be secured on the spot, such as making a yield check on corn or taping several steers to estimate their weight.
5. *What changes need to be made at this time?* This raises the question of what procedures need to be applied to avoid the undesirable outcomes which have been identified. In the example of the steers, the protein level of the concentrate ration may be decided upon as the limiting factor. Questions of this nature are important not only in determining what changes need to be made in farm practice, but also to develop in the student habits of sound decision making and of rational thinking.
6. *What will be the probable cost and return in making the change in practice?* This is an important question in the evaluation process since it deals with the economic realities of making changes in the situation. Often the costs of making a change may be less than the student has realized and the returns may be considerably greater. These costs and returns should be calculated on the spot if possible and on the basis of the actual situation. In the example of the steers, the cost of an increase of 2% in the protein ration can be computed and the cost for this particular lot of cattle can be compared with the increased returns which may be expected as a result of the change in their ration.
7. *What further adjustments and procedures will be needed before the final goal can be achieved?* This question should cause the student to look forward to the completion of the operation which is being discussed and to discuss additional practices which will contribute to the success of his enterprise. During such discussion the steer feeder, used as our example, may give thought to time and place of marketing or length of feeding period as well as feeding practices.

This informal approach to appraisal of farm practice has been long used by teachers who were skilled in human relations. It has the advantage of taking the teacher out of the role of the "supervisor" who "tells" the student what to do and places him in a more appropriate role as a guide who raises questions to which both student and teacher seek to find appropriate answers. The results are important in that the student not only learns to apply the teaching of his adult education program but also he develops habits of evaluating his own efforts. The fact that he develops methodical habits of evaluating his own efforts often contributes materially to his future success.

ADVANTAGES OF EVALUATION

An organized and continuing plan of evaluation offers many advantages both to the teacher and to the students. The local leader is likely to develop a better understanding of his program if he develops, in writing, the objectives which he is seeking to attain. This also holds true for members of the class for they too develop a greater understanding of their program through helping define its goals.

An organized program of evaluation gives the teacher or leader a sense of security and accomplishment through knowing what is expected of him in his adult education program. He can take pride in his work because it measures up to certain standards which have been set by himself, and members of his class, and which are acceptable to the other teachers or agents in his district or region and state.

The findings of evaluation serve as a sound basis for future program development. One of the major functions of evaluation is to help make decisions among alternatives. The teacher or leader of an adult group makes decisions such as how much time should be spent in adult education, what clientele groups should be served, what kind of knowledge and skills should be taught, when and where courses should be offered, and how much time should be spent in the classroom as compared to time spent on field trips and in individualized instruction. Evaluation provides a basis for the soundest possible decisions on such problems.

Organized evaluation also provides an improved basis for the use of supervisory assistance. This is particularly true if administrative and supervisory personnel are involved in the evaluation process.

Adult students need to learn how to use the process of evaluation as they make day to day decisions. Much of the teaching of business managers is concerned with helping them to evaluate what they are doing as a basis for making further improvement. The success of farm business planning courses lies largely in the fact that through keeping accurate business records, farm families are provided an opportunity to evaluate the financial progress of their business in terms of goals which they are seeking to achieve. Changes in the farm business are made with a clear understanding that such change is necessary to achieve the goal of improved farm income.

Another advantage in evaluation comes from the greater public understanding which is developed—particularly when evaluation extends beyond the participants in the program. As representatives of the school and the community are involved in evaluation, they become familiar with the purposes and accomplishments of the adult program and they spread the word to others.

Finally, adult education is enhanced by evaluation because the participants come to see the program as something they have a share in, rather than as the exclusive property of the extension agent or the teacher.

Many writers support the increasing need for evaluation of adult education programs. Bundy (2) says that most teachers of vocational agriculture should devote more time to evaluating their adult farmer programs. He believes that evaluation should be made in numbers of persons who participate, the regularity of their participation, and the practices put into operation on their farms as a result of the instruction. Bundy states that the methods used should be evaluated so that improvement may be made, and that additional instructor time must be available in order for the instructor to assist farmers in bringing about changes in farm business management. He believes that final evaluation of the effectiveness of vocational agriculture programs for farmers should rest upon the improvement in the financial condition of the farm operators, which can be determined only by the use of farm management and farm record keeping programs.

Procedures and Techniques

The whole movement of evaluation of agricultural education programs in local communities seems to have become bogged down by the very procedures which have been recommended for such evaluations. Most states have developed instruments for the appraisal of local agricultural education programs including adult education. However, these instruments are so voluminous and the evaluations become so time consuming, require so many records, and so much study, that only a few persons understand them and only a few teachers have found the time to make use of them in evaluating their programs. What is needed, apparently, are simpler instruments and more practical evaluation procedures which can be combined with a variety of informal evaluation procedures commonly used by teachers of adults. This section outlines some plans which can be followed by the teacher or leader who seeks to evaluate his own program.

In making these plans, the teacher should begin by evaluating some aspects of the program using his own initiative and his own procedures.

In making such an evaluation, a combination of "self" and "outside" evaluation would be most effective. When we evaluate ourselves we accept our own judgments, but we often feel more secure when we have the judgments of others to reinforce our convictions. These "others" may include professional persons, such as administrators and supervisors who have had the opportunity to see other programs as a basis of comparison. In such an evaluation it is assumed that as many as possible of those affected by the program will share in its evaluation. This means that the adult education program will be evaluated by farmers, businessmen, school administrators, and other community leaders. If such an evaluation is to be made, it must be economical of the time and resources of those who participate, and usually this means that the instruments must be simple and yet precise and effective.

PROCEDURES IN EVALUATING LOCAL PROGRAMS

The following is a 12-step procedure appropriate for conducting an organized evaluation of an adult education program. This procedure has demonstrated its worth in a number of different schools and communities during the course of evaluations of vocational agriculture programs. The procedures can be adjusted to different situations, depending upon the amount of time available and the extensiveness of the program to be evaluated.

The following are the major steps arranged in chronological order:

1. The local teacher or leader should initiate the evaluation explaining its value and the major procedures which will be involved. This explanation can serve as a basis for securing the approval of the superintendent of the schools, the director of the local vocational education program, and the supervisor of agricultural education.
2. The local administrator, upon the teacher's recommendation, should appoint a professional evaluation team. Persons on this team should include at least one farmer, one member of the local administrative team, one principal or guidance director, one or two neighboring teachers and a supervisor of agricultural education. This evaluation team should assume the responsibility for planning the balance of the evaluation, for conducting it, for making the final evaluation, and for preparing the written report.
3. A convenient date should be set for the evaluation by the professional evaluation team.
4. The evaluation team should determine the major sources of information needed for the evaluation and develop the needed instruments for securing and recording this data.

5. The evaluation team should name an advisory panel of six to twelve persons to assist in the evaluation. These persons should include participants of the adult education program as well as local administrators, local businessmen and others affected by the program. The primary function of this advisory panel is to make judgments on the outcomes of the program. They should be persons intimately connected with the program and have the ability to make fair and honest judgments of its progress.
6. The teacher or teachers should collect and assemble needed information and have it available for use in the evaluation.
7. The evaluation team should direct the activities of a one-day evaluation session including an evening dinner meeting. If time is more limited, a half-day and evening session may be sufficient. Ordinarily the evaluation team should begin their work by becoming acquainted with the program through examining completed instruments and evidence previously developed.
8. At some time during the day or evening, the advisory panel should meet with the evaluation team. This advisory panel usually requires one to two hours for their discussion of the on-going adult program. One of the characteristics of such groups is that they are at first inclined to be overly protective of their teachers. However, as they learn the purposes of the evaluation, and as they become acquainted with their responsibilities, they begin to react with more freedom.
9. Adroit and skillful questioning on the part of the evaluation team is one of the keys to securing judgments of the panel on the outcomes of the adult education program. A series of carefully structured questions is necessary to secure the judgments and opinions of the advisory panel. The first question should consider accomplishments of the program of adult education during the past few years. Such questions give the panel an opportunity to recall significant events and to recognize the efforts of their local leaders. Another series of questions should consider changes which have occurred in agriculture and in the community schools within the past few years. This gives the committee a chance to consider some of the changes in agriculture and in education which are important to the development of a program. The final questions can then deal with changes and improvements which need to be made in this adult education program. Usually, general questions such as those above are more effective in eliciting judgments of adults on the effectiveness of the program than are more direct questions.
10. The evaluation committee, after thanking and dismissing the panel, should complete the evaluation. A written report of the

evaluation should be prepared and submitted to the administrator of the school and/or to other appropriate individuals.

11. A follow-up of the conference should be held about six months later to review new developments which are proposed as a result of the evaluation.

INSTRUMENTS FOR SUPPLEMENTARY AND INFORMAL EVALUATION

Informal evaluation of various aspects and procedures of the program of adult education is a valuable supplement to the complete evaluation which was described earlier.

The use of a variety of procedures for evaluating an adult farmer course is described by John F. Cassidy, who has taught adult farmer classes for 28 years at Creedmoor, North Carolina. Cassidy's program utilizes a planning committee, informal feedback from members, printed evaluation instruments and community studies. Cassidy describes his program as follows:

> We use an advisory council composed of nine members equally distributed throughout the school district to act as a sounding board as to the needs and desires of the farmers of the community. This group acts as a planning and evaluating committee during their semi-annual meetings, one for the summer program and one for the winter program. The summer program is planned around tours to outstanding farms, to experiment stations and a county-wide tour. The winter program is composed of special courses and a general subjects course.
>
> Early in the fall we have an organizational supper meeting, sponsored by the local bank, at which each person is given an opportunity to register for the course or courses he desires. To warrant the presentation of a course, registration must number at least ten.
>
> In the spring we complete our organized class meetings with a recognition barbecue supper at which we also summarize our accomplishments for the year and give members an opportunity to make suggestions for next year's program. These suggestions are given careful consideration at our next advisory council meeting in planning the offerings for the following year.
>
> Our T.V. courses are made up of a one-half hour T.V. presentation followed by a one and one-half hour classroom discussion. We are furnished printed evaluation sheets, by the state department of education, which each participating student checks at the completion of the course. This sheet serves our purpose of evaluation. It is also used by the state department of education in making adjustments in courses to follow.
>
> We have been fortunate for the past five years to have had excellent student teachers from North Carolina State University who have

made comprehensive community studies as a requirement of a rural sociology course. These studies have provided us a continuing evaluation of our program on many points we may have otherwise overlooked.

Essential Characteristics of Evaluative Instruments. An instrument appropriate for evaluating an adult education program generally consists of a series of shorter instruments each relating to one of the major objectives of the program. Using this approach an evaluation instrument might include sections on course of study, teaching methods, on-farm teaching, enrollment, changes in farming and civic and social activities.

Each of the major objectives becomes the basis for a short instrument, perhaps a page in length, which should be developed by the particular group who were evaluating this program. Generally, each of these instruments should contain the following six types of information which would be used as a guide for evaluating each individual phase of the program.

1. A *guideline* which would delineate the contributions of the particular objective.
2. *Facts and figures* regarding this aspect of the program for the current year. These facts and figures are objective in nature and should consist of only a few key indicators rather than all the information which might be compiled.
3. *Selected trends* for the past three to five years should be recorded. Trends provide a further look at what has happened in the program and may be used in the forecast of what may happen in the future.
4. *Judgments of results* or outcomes in this aspect of the program should be secured. If we limit the evaluation to ways and means as we tend to when we examine figures and trends we fail to include the outcomes which people are most concerned with.
5. A *simple and understandable rating scale* for summarizing the evaluation of the area.
6. *Provision for recommendations* for further improvement.

The instrument shown on page 209 is an example of one which was used for evaluating a young farmer program (13). Although designed as a part of a larger instrument for making an evaluation of a total program, it illustrates each of the points which have been made and is a desirable instrument in that it is brief, to the point, and provides a variety of facts and judgments for an evaluation.

Finally, the process of developing your own evaluation instrument may be as important as the instrument itself. Instruments of evaluation have maximum meaning for those who hammer them out and put them together. In developing instruments for individual programs each teacher or leader has the responsibility of involving in their development all of those who will use them.

Young Farmer Program

Guiding Statement: A special educational program for young farmers should be offered which provides for continuation of their education regarding efficiency of production, farm business planning and establishment in agricultural careers. Opportunity to develop further social, recreational, and civic understandings and appreciations should also be afforded.

1. *Some Facts and Figures*

	This Year
List significant farming changes made by students resulting from organized instruction	______
Number of former high school vocational agriculture students now enrolled in the Young Farmer Program	______
Number of field trips	______
Number of sessions devoted to farm business planning and analysis	______
Number of sessions in which resource persons were used	______
Number of years of affiliation with the State Young Farmer Association	______

2. *Selected Trends*

	2 Yrs. Ago	*Last Year*	*This Year*
Number of instructional meetings held	______	______	______
Average attendance at instructional meetings	______	______	______
Total enrollment	______	______	______
Number of teacher visits per student per year	______	______	______
Number of teacher visits made to prospective members per year	______	______	______

3. *Outcomes or Results*
Use local interviews, observations, and discussions as a basis for your appraisal of results. Evaluation Scale: 5-Excellent; 4-Good; 3-Average; 2-Fair; 1-Poor.

5 4 3 2 1

. . . . . 1. The instructional meetings are interesting and useful to those enrolled.

. . . . . 2. The members are making desirable changes in their farming businesses as a result of the course.

. . . . . 3. The members are developing further competency in social, recreational, civic, and leadership responsibilities.

. . . . . 4. The members are becoming successfully established in farming.

4. *Evaluation of Area*

To what extent does the Young Farmer Program contribute to the success of its members in becoming established in agricultural careers, in improving their farming operations and in improved

5 4 3 2 1 rural living.

— Use back of page for other major accomplishments in this area and suggested improvements —

Selected Evaluation Instruments. Following are examples of instruments which may be used for certain on-the-spot evaluations. It is hoped that they may help serve as patterns, for generally such instruments are of greater value when they are developed by the teacher or the group who will be making the evaluation. In developing such instruments, adult class members usually gain a greater understanding of the semantics of the criteria.

The evaluation of classroom teaching is of special interest to the local leader of the adult education program. The "Guide for Evaluating Lesson Plans for Teaching Farmers" is an aid for self-evaluation on the part of the teacher. With some revision it might also be distributed to members of the class in order to get their appraisal.

Another instrument of somewhat broader scope is entitled "Suggestions for Improving the Course." As the title implies, this form provides a basis for a comprehensive evaluation involving a series of meetings. One of the interesting aspects is that the two sides of this instrument provide space for the participants to indicate strengths on the one hand and weaknesses on the other hand. Such an instrument has the advantage of securing both strengths and weaknesses, while many instruments list only the desirable factors involved.

The instrument "How Am I Doing as a Committee Leader or Chairman," (1) is adapted for use by the instructor or by the participants and provides an opportunity for some evaluation of the human relations aspects of the teacher as he works with his adult education group. The emphasis of this instrument is upon democratic procedure characterized by informality, recognition of leadership from the group, and of a problem-centered, thought-provoking approach to teaching. The use of such an instrument is not only helpful as a basis for self-evaluation by the teacher but also can give students a better insight as to what the teacher is attempting to do through his discussion.

The most comprehensive instrument which has been included in this group is "How Good is My Adult Program." This instrument has been used by many teachers of adult education classes in Ohio, and it has the advantage of serving as an instrument for evaluation at the end of a course or series of meetings.

A Guide for Evaluating
Lesson Plans for Teaching Farmers

Clear and specific objectives have been identified.

A challenging introduction is used to start the discussion.

The problem is timely, useful, of proper scope, and is clearly stated.

A logical plan for solution of the problem is proposed. This plan includes the problem situation, factors and possibilities.

Provision is made for students to secure facts from a variety of sources.

Appropriate factual information has been assembled by the teacher sufficient to support the conclusion or decision of the class.

Teaching aids and other techniques for developing interest and providing student experience are included.

Procedures and questions for promoting evaluation of facts and group thinking are pre-determined.

A practical conclusion or decision can be reached by the group at the conclusion of the presentation.

Suggestions for Improving the Course

Your suggestions can be of real help in making further improvements in this course. Will you list briefly what you consider the strengths and weaknesses of the organization of this course as well as the teaching procedures which were used. Please make your suggestions under the headings below.

Do *NOT* sign your name.

1. Stimulation of *thinking* and *understanding*.

   This worked / This didn't seem to work

   ____

   ____

   ____

2. *Degree of interest* developed during the course.

   These things helped / These hindered

   ____

   ____

   ____

3. Appropriateness of the bulletins and mimeographs which were provided.

   Desirable / Undesirable

   ____

   ____

4. *Usefulness* of the *subject matter* to you.

   More emphasis should be given / Less emphasis should be given

   ____

   ____

   ____

5. In comparison with other courses taken to date, rate this course by underscoring one. Excellent Good Average Poor

6. Which of the sessions did you like best.

   ____

   ____

   ____

   ____

   ____

7. On the back of this sheet, list any other suggestions which might have made this course of greater value to you.

How Am I Doing?

As a Committee Leader or Chairman,
A Checklist of Favorable Symptoms!

________ Members address me no more formally than others in group.

________ Members frequently express real feelings.

________ Group starts itself at beginning of each meeting.

________ Sometimes members openly disagree with me.

________ Members address their remarks to each other rather than to me.

________ Group has a tendency to want to remain after the time limit has passed.

________ Group makes decisions without depending on me as the final judge.

________ Members seem to know what goals they seek.

________ Members speak up without asking for my permission.

________ Members do not count on me alone to handle "problem members."

________ "Bright ideas" originate with many members of the group.

________ Different individuals frequently lead the group's thinking, discussion and procedure.

________ Members often accept insights and information from other members.

________ There is an absence of hostility toward me.

________ Members draw out and question each other to better understand their contributions.

Source: How to Lead Discussions, (Leadership Pamphlets #1), Adult Education Association of U. S. A., 1955.

Evaluating the Chairman

As a Leader or Chairman, I Will Evaluate Myself!

Certain qualities in a chairman or leader promote good committee meetings. Here are a few for you to test yourself with. Score yourself 3 points for each "always," 2 for "sometimes," and 1 for "never." A good discussion leader will score 20 to 30 points.

	Always	Sometimes	Never
Do I get all possible information to the members before the meeting?	______	______	______
Have I checked arrangements for the meeting, such as heating, lighting, seating, etc.?	______	______	______
Do I have agenda worked out, estimating the time likely to be required for each segment?	______	______	______
Am I tolerant of members with whom I disagree?	______	______	______
Do I use questions during the discussion to encourage participation?	______	______	______
Can I handle an argument without getting personally involved?	______	______	______
Do I try to get the less vocal members to participate?	______	______	______
Am I aware of the importance of adequate information for a discussion?	______	______	______
Do I come to meetings prepared for a clear introduction of the subject, aimed at clarifying goals and promoting informality?	______	______	______
Do I listen to detect when members are in agreement on an issue, or when they are getting off the subject?	______	______	______

My score .... ______

Source: The Chairman's Job, Special Circular #44, Penna., State University College of Agriculture and Extension Service, University Park, Pa.

How Good Is My Adult Program?

Criteria	Rating (check one)		
	To a Great Extent	To Some Extent	Not at All
1. Local community members are aware of the need for adult education.			
2. Community members are helping to determine the needs of adults in the community for education in agriculture.			
3. The program is based on the educational needs in agriculture of the people of the community.			
4. Education in Vocational Agriculture is provided for all those who need it.			
5. The participants in the program are helping to plan for, conduct and evaluate it.			
6. A wide variety of approaches is used to meet the needs of and interest all groups in the community.			
7. The approaches being used resulted from following the "steps" in successful program planning.			
8. Schedules and methods are designed or adjusted to meet the needs of the participants.			
9. Adequate funds are provided for developing and conducting the program.			
10. Adequate time is provided for the vocational agriculture teachers to provide leadership for planning, conducting and evaluating the programs.			

How Good Is My Adult Program?

Criteria (cont.)	Rating (check one)		
	To a Great Extent	To Some Extent	Not at All
11. Leadership of the program is provided by the school's chief administrative officer.			
12. The program is coordinated with other educational offerings by the school.			
13. Continuous research is being conducted to give direction for changes for improvement of the program.			
14. Objectives for the various groups have been determined and are being met.			
15. A system of continuous evaluation is built into the program plan.			
16.			
17.			

COMMENTS:

____

____

____

____

____

____

____

____

Date of Evaluation ____________________________

Evaluators(s) ________________________________

While these instruments offer some structured approaches to evaluation, there should be frank discussion between members and the teacher at the conclusion of meetings, while on farm visits, and on similar occasions. Such informal conversations may be even more revealing than some of the instruments which have been suggested above.

As an example of a teacher's approach to evaluation of his adult education program, Robert Brinson of Michigantown, Indiana says, "Pretests of a few questions are sometimes desirable to start discussions but written tests are not well received by farmers."

Brinson, who has taught young farmer classes for nine years, says his program ordinarily includes 10 meetings on farm management, 10 tour meetings and 10 farm mechanics meetings. Brinson supports the value of securing informal oral evaluations of the program as it develops. He says, "I believe the best evaluation takes place on farm visits when we talk about our recent meetings and how to improve them. This means an oral evaluation with the individual. Some of this type of evaluation is made in the 'bull sessions' which follow our meetings."

Brison believes continued attendance is the best single indicator of the success of his adult education program. He says, "To me, the most important question is 'Does my program help my agricultural people become more proficient in their area work?' Continued attendance and growing attendance are excellent indicators to me."

Conclusion

Evaluation is an educational process which can help to determine whether the objectives of the program of adult education are being met. No one method is foolproof, nor are the standards infallible. Teachers and leaders of continuing education programs in agriculture should evaluate their program using a wide variety of evaluative procedures.

Practical steps which can help to integrate evaluation into the young and adult farmer program include periodic organized appraisals supplemented by a variety of informal approaches. Planning committees should be utilized to help gather and interpret factual information regarding the program and should serve as a method of informing the teacher of student criticism of the program. Interviewing farmers, informally or through the use of checklists, to ascertain their interests and their problems is another approach to evaluation which should not be neglected. To be most effective, evaluation should be built into the complete program of adult and young farmer education on a continuing basis. The entire class and planning commitee should be included in the evaluation process in order to guard against limited or biased approaches and to give the class members a feeling of involvement in the direction and outcome of the program.

Finally, evaluation should become a part of the teaching process. It offers a means of making farm visitation more effective and helps members to make use of the process in decision making and in learning to become better farmers and citizens.

REFERENCES

1. Adult Education Association of the U. S. A., "How to Lead Discussions," Leadership Pamphlet No. 1, 1955.
2. Bundy, Clarence E., "March Is a Good Time for Evaluating Adult Farmer Programs," *Teacher Education.* Columbus: The Ohio State University, Vol. 5, No. 8 (March, 1963).
3. Good, Carter V. (ed.), *Dictionary of Education.* New York: McGraw-Hill Book Company, 1945.
4. Humphrey, Carl M., "Adult Farmer Education Deserves Evaluation," *The Agricultural Education Magazine,* Vol. 35, No. 5 (1962).
5. Knox, Allen B., "Developing an Evaluation Plan in Adult Farmer Education," *The Agricultural Education Magazine,* Vol. 35, No. 5 (1962).
6. Office of Education, "An Evaluation for Local Program to Vocational Education in Agriculture," Federal Security Agency, Vocational Bulletin 200240.
7. Persons, Edgar A., *et al,* "Investments in Education for Farmers," *Research Report on Project #42765.* St. Paul: University of Minnesota, January, 1968.
8. Stufflebeam, Daniel L., *Evaluation as Enlightenment for Decision Making.* Columbus: The Ohio State University, Evaluation Center, College of Education, 1968.
9. Sutherland, Sidney S., "Objectives and Evaluation in Vocational Agriculture," *Evaluation and Program Planning in Agricultural Educaation.* Columbus: The Center for Vocational and Technical Education, The Ohio State University (1966), pp. 13-18.
10. U. S. Office of Education and the American Vocational Association, *Objectives for Vocational and Technical Education in Agriculture.* Prepared by a Joint Committee of the U. S. Office of Education and the American Vocational Association. Washington, D. C.: U. S. Government Printing Office, 1965.

11. USDA-NASULGC Study Committee on Cooperative Extension, *A People and a Spirit.* A report of the Joint USDA-NASULGC Study Committee on Cooperative Extension. Fort Collins, Colo.: Coloradio State University, November, 1968.

12. Verner, Coolie, Booth and Allen. *Material Selected from "Adult Education."* New York: Center for Applied Research, Inc., 1964.

13. Woodin, Ralph J., *Appraising the Vocational Agriculture Program.* Columbus: Department of Agricultural Education, The Ohio State University, June, 1970.

INDEX